THE ROBOT AND AUTOMATION ALMANAC 2018

 THE FUTURIST INSTITUTE

Edited by Jason Schenker
Chairman of The Futurist Institute

For my automation and robotics colleagues.

THE FUTURIST INSTITUTE

CONTENTS

PREFACE   9

INTRODUCTION: THE FUTURE HISTORY OF ROBOTS   13
Jason Schenker

SECTION ONE: ROBOTS YOU SEE   17

OVERVIEW: ROBOTS YOU SEE   19
Jason Schenker

WAITING ON ROBOTS   21
Will Allen

THE NEW WAVE OF ROBOTS   27
Jeff Burnstein

THE RISE OF SERVICE ROBOTS   33
Steve Cousins

**ROBOTICS AND AUTOMATION IN CONSTRUCTION
GAIN MOMENTUM**   39
Kaleb Steinhauer

HYPERLOOP AND THE FUTURE OF TRANSPORTATION   45
Holly McNamara

ROBOTS IN HOSPITALS AND PHARMACIES   51
Stanislaw Radominski

ROBOTS AND AUTOMATION ON THE ROAD   57
Craig Fuller

CHATBOTS THAT TALK   61
Lorenzo Carver

INVESTING IN ROBOTS   65
Jeremie Capron

CONTENTS

ROBOTS IN ENERGY   71
John Gibson

ROBOTS IN FINANCE   79
Jason Schenker

BUILDING THE RIGHT ROBOTS   85
Bruce Welty

SECTION TWO: ROBOTS YOU DON'T SEE   91

OVERVIEW: ROBOTS YOU DON'T SEE   93
Jason Schenker

ROBOTS AND AUTOMATION ARE NOT OPTIONAL   97
Michael Walton

ROBOTICS IN CHINA — THE NEXT FRONTIER?   101
Henrik I. Christensen

THE FUTURE IS BUILT UPON THE PAST   105
Jayesh Mehta

THE IMPACT OF AUTOMATION ON THE SUPPLY CHAIN   111
Daniel Stanton

**ROBOTICS IN THE WAREHOUSE: THE AUTONOMOUS
TRANSPORTATION REVOLUTION COMES INDOORS**   119
Lou Micheletto

**SUPPLY CHAIN AUTOMATION:
HUMANS WILL ALWAYS BE ON THE LOOP**   125
Rob Handfield

SOFTWARE-DEFINED AUTOMATION   135
Louis Borders

CONTENTS

COZYING UP TO THE LAST MILE
David Schwebel

141

**ACCEPTANCE AND INDUSTRY 4.0
DRIVE ENTERPRISE DEPLOYMENTS**
Laura McConney

149

**THE NEED TO OVERCOME INTEROPERABILITY
AND LABOR REPLACEMENT FEARS**
Daniel Theobold

155

REDEFINING GROWTH
Fred van Beuningen

161

ABOUT THE FUTURIST INSTITUTE

167

ABOUT THE EDITOR

169

ABOUT THE PUBLISHER

173

DISCLAIMERS

175

FROM THE FUTURIST INSTITUTE

On behalf of The Futurist Institute, I want to thank all of the contributors to *The Robot and Automation Almanac - 2018*. We have such an amazing cohort of contributors in the fields of robots and automation. I am truly humbled by the list of authors.

Thank you to Will Allen, Fred van Beuningen, Louis Borders, Jeff Burnstein, Jeremie Capron, Lorenzo Carver, Henrik Christensen, Steve Cousins, Craig Fuller, John Gibson, Rob Handfield, Holly McNamara, Jayesh Mehta, Laura McConney, Lou Micheletto, Stanislaw Radominski, David Schwebel, Daniel Stanton, Kaleb Steinhauer, Daniel Theobald, Michael Walton, and Bruce Welty.

The contributions of these leaders will help shape the vision for robots and automation in the year to come — and beyond. Their words and expectations for "the big thing" in the year ahead have created an important vision of the future. And without them, this book would never have happened. This has been an exciting project for me, and I believe you will also find the visions of our authors for the year ahead to be exciting!

I also want to thank parties who have not contributed to this work but who have helped out by providing recommendations for potential contributors, including MHI — the Material Handling Industry group — with which my financial market research firm, Prestige Economics, has a long-standing relationship. I also need to thank Martin Buehler, the executive director of robotics at Walt Disney Imagineering in California. He introduced me to several of the authors in this work. They are experts, and without Martin's introductions, this book would not be as rich as it is.

Finally, I need to thank the individuals who have provided support and feedback to this project. Kerry Ellis did an amazing job producing the cover for *The Robot and Automation Almanac*. And Nawfal Patel and my other colleagues at Prestige Economics and The Futurist Institute helped me bring this book to fruition.

It is also my personal pleasure to have been involved in this undertaking. The robotics and automation industries are still very fluid, but they are becoming more important. And it is with great pleasure that The Futurist Institute is able to support the companies in these industries with this piece of knowledge capital. It is also our pleasure to work alongside the critical long-established thought leadership and conferences that have been so important for the development of robotics and automation, including A3 and the RoboBusiness conference.

The Robot and Automation Almanac represents a written gathering of thought leaders, but RoboBusiness is an amazing in-person experience. It has been formative for my own transition from economist to futurist — and for the very founding of The Futurist Institute.

The Futurist Institute was founded to help analysts and economists become futurists. With the inaugural edition of *The Robot and Automation Almanac* in 2018, we believe The Futurist Institute has created an invaluable tool for those professionals, individuals, and investors who seek to understand the implications of robots and automation for their personal, professional, and investing lives.

The transition from the information age into the automation age is underway. Thank you for being a part of this dynamic shift.

Welcome to the future! ~

Jason Schenker
Chairman of The Futurist Institute
Editor of *The Robot and Automation Almanac*

The Future History of Robots

The history of robots is being written now.

In this book.

When people look back a decade from now, they will want to know how robots and automation evolved. They will want to understand how robots emerged from the factories. They will wish they had known more when it was happening.

But rather than wait for a historian to document the future, *The Robot and Automation Almanac* captures perspectives on imminent pivotal changes *now,* in the tectonic transition from the information age to the automation age.

We are standing on the edge of the age of automation. What's next for robots and automation in the year ahead? This is the question that almost two dozen contributors have answered in this book.

Robots and automation received a lot of media attention in 2017. But this visibility and emergence in the technological *Zeitgeist* was a shadow of what is likely to happen in 2018. Our authors have presented a view of a world in which robots and automation will make more rapidly accelerating progress. It may not be the year in which robots are a raging river of technological disruption in our daily lives, but 2018 is certain to be a year in which what had been a trickle will be emerging as a babbling brook. Where there was a little activity before, there will be much more in the year ahead.

In the essays in this book, you will see a common thread among the different authors: 2018 is a year of ramping up activity. While it may not be the year of full escape velocity, it is going to be a big year. We will likely look back at 2018 as a critical year of transition, of momentum, of activity, of investment, and of technological potential.

We have divided *The Robot and Automation Almanac - 2018* into two parts: "Robots You See" and "Robots You Don't See."

Robots You See

In the year ahead, robots are likely to emerge from the factory, as Will Allen, Jeff Burnstein, Steve Cousins, and others have noted in their essays. These essays are critical parts of the "Robots You See" section of this book. We have included a number of other topics, including transportation and energy, in this section of the book. And we have included an essay from Jeremie Capron, the head of research for the largest robotics exchange-traded fund in the world, ROBO Global, to comment on investments in the space.

As robot and automation technology become more visible, interest in the space is likely to garner more attention from investors as well, from venture capital funds and angel investors to retail investors who want to be able to own a piece of the robot and automation story.

Robots You Don't See

While we are likely to see more robots in the year ahead, technological developments that allow for a greater integration of software and hardware are likely to accelerate as well. They may go unnoticed by people who are not professionally entrenched in the world of robots and automation. But they will be significant. This is the subject of Michael Walton's essay in the section of this book that is focused on "Robots You Don't See."

Of course, our authors have also addressed some of the politics of robots, the ascendance of Chinese robots, and the importance of robots and automation in the supply chain. After all, robots and automated solutions are *conditio sine qua non* to fulfilling the promise of e-commerce.

The history of robots and automation is being written now. And while consumers will see the difference a few years out, this critical emerging trend is marching forward now. Even if not everyone can see it. ~

Jason Schenker
Chairman of The Futurist Institute
Editor of *The Robot and Automation Almanac*

Robots You See

 THE FUTURIST INSTITUTE

Robots You See

Jason Schenker

- Chairman of The Futurist Institute -

This is the year you will begin seeing robots in your day-to-day life.

They won't be everywhere, but our authors have made a compelling case that 2018 will be a year of transition. In 2017, you heard about robots — and you might have even read my book *Jobs for Robots*. In 2018, the robots will begin to more visibly arrive on the scene. And it will just be the beginning.

Robots will not only become more visible, but they will also become increasingly important for your personal, professional, and investing lives. In this section of *The Robot and Automation Almanac*, we have pulled together essays on robots you are likely to see and be exposed to in the year ahead.

This section includes robots you might see in hotels, at the pharmacy, on the roads, and in the oil patch. Plus, this section includes an essay about where we want to see robots: working in our homes and cleaning our bathrooms, as Will Allen of HP Labs is quick to note.

This section also includes essays about investing in robots and robots in finance, as well as the coming wave of chatbots that could affect your day-to-day activities in 2018.

We conclude this section with an essay by Bruce Welty about the need for more practical robot design. While we are likely to see more robots in the world, they are unlikely to be the anthropomorphic robots of Hollywood blockbusters. You are unlikely to encounter robots that look like humans and climb trees.

But they will be out there in 2018. And they will be making your life easier. ~

Jason Schenker
Chairman of The Futurist Institute
Editor of *The Robot and Automation Almanac*

Waiting for Robots

Will Allen
- HP Labs -

According to Nielsen, there are about 120 million TV homes in the United States for the 2017-2018 viewing season. Neil Armstrong walked on the moon in 1969...almost 50 years ago. I find it astonishing that useful robots are not as common in homes as televisions.

When will this situation change? When will this wave of new technology sweep over our planet? It turns out technology "revolutions" are spread out over periods of years. The recent explosive wave of smartphone technology actually started over two decades ago when shoe-sized first-generation mobile phones began appearing intermittently in our lives.

I suggest we watch the set of tools available to those researching and developing solutions based on robots. Tools are a fundamental enabler, and investment in tools will increase as more and more robots are deployed. Tool capability will be an important predictor of prominence of non-industrial robotics as an industry. Who wouldn't be happy with a personal C-3PO valet? "3PO, the Nguyens are coming over for dinner. Please clean the guest bathroom and toilet."

This wave of new technology will not start with anthropomorphic talking robots akin to C-3PO from *Star Wars*. Some of today's most advanced robotic systems participate in the DARPA Robotics Challenge. From even a quick look at robots participating in the 2017 DARPA Challenge, it's easy to see we are some years away from affordable mass production of useful human-sized robots capable of walking around on two legs.

C-3PO is a real character in cinema; in our reality, it's likely robots with less exciting physical capabilities will be the pioneers. Watch for emerging robots more akin to R2-D2: computerized machines delightfully rolling around conducting occasional modest physical interactions with objects in our environment.

Autonomous execution of complex tasks is one goal for designers of robotic systems. It turns out robotic systems not capable of achieving that goal can be very valuable.

Sometimes tasks accomplished by performing only simple operations have great value. Tasks with lower-performance difficulty levels are delivered by less complex, hence less expensive, robotic systems. Here's how it works...

> Mario is a busy work-from-home dad. This afternoon he has to help his daughter with calculus homework, finish a bid for his wedding planner job, and pick up toys in the playroom. Mario doesn't have time to get everything done — he's overrun. Mario's friend Pat stops by and sees Mario is really busy, so Pat helps out by picking up the playroom toys. These efforts save Mario twenty minutes, which gives him time to get his bid finished and help his daughter get started on her calculus homework.

The key here is an asymmetry: when someone is overrun with tasks, saving one minute on the simplest task can have the same value as saving one minute on the most complex task. Robots will appear with increasing frequency in our homes, in retail outlets, and in our offices. First, we will see simpler robots with modest capabilities for mobility and environmental interaction. These robots frequently will supply value by performing simple tasks or supplying elementary assistance to individuals working on complex tasks.

Today one sees that most significant impediments to affordable robotic solutions have been addressed. High-density energy storage is in our hands: just look at cordless hand tools and drones.

Powerful high-value computing and sensor resources are not hard to come by. Machine learning is advancing at a rapid rate every day, and multiple hardware acceleration platforms are available. Integration of affordable electromechanical systems isn't an impediment: incredibly complex digital printers are very affordable. The open source Robot Operating System is a highly accessible platform for research and development of interesting robotic solutions.

The pieces are all on the table to fill the world with valuable robotic solutions. Cost-effective components exist up and down the technology stack. Security is critical, and robotic systems will take advantage of solutions already in enterprise and consumer applications protecting printers and personal computing systems. Global scale to manufacture and service electromechanical devices already exists.

What's missing? Refined tools including a rich set of software-based components. In the 1990s, publishing a blog would have required a lot of web programming just to get started. Today, in 30 minutes, virtually anyone can have a professional-looking blog up, simply by using any one of a dozen free services with drag-and-drop interfaces. Today's developers of robotic solutions don't enjoy a rich set of mature tools. Their world is somewhat analogous to the universe facing bloggers twenty years ago. Products can be created, but the time and effort are large relative to the value of the result.

When will robots be as common as televisions? Not for some years, but...I believe 2018 will be a pivotal year for non-industrial robotics. The pivot will come in the form of continued growth of tools aiding those creating and deploying robotic solutions. This will enable many new offerings to enter the marketplace, both from entrepreneurs and from established technology players. Some of these solutions will not be very valuable, and some will be compelling. A growing installed base of useful robots is a magnet for investment by providers of tools serving creators of those solutions.

Demand for non-industrial robots will only increase. HP cited an aging population demographic as a key megatrend for two consecutive years. According to the U.S. Social Security Administration, in 2013, for every retired American, there existed 2.8 active workers. Demographer Jonathan Last predicts that in 2040, each retired American will have only 2.0 active workers making economic contributions to help support them. As our parents and loved ones grow old, we must find more efficient ways to connect with them and care for them.

I have personally experienced increased value using mobilized telepresence technology to connect with elder family members. The results were amazing, with seniors changing habits after the introduction of a mobile telepresence device into their household. Even though nobody was physically traveling, the conversation shifted from "What time will you call us Sunday?" to "When are you coming over Sunday?"

This is not simple task automation — it's creating something that wasn't possible in the past. Low-mechanical-complexity robots can connect people in ways that were previously impossible. Mechanically-unsophisticated social robots are proven to help people improve habits, and social robots may reduce loneliness. Robots capable of reducing human time spent on low-value tasks such as making sure doors are locked, or finding toys, will supply busy people precious minutes to spend on fulfilling hobbies, personal education, and helping aging family members.

The universe of non-industrial robots is on the cusp of becoming significant — better tools fostering better solutions fostering increased investment in tools. Components will increase in value, many exponentially. Demand is already large, though latent, and megatrends will grow demand significantly for decades.

Watch the capabilities of the toolsets and software components available to those who are researching and developing robotic solutions. The collective state of those tools is an excellent harbinger of an exciting new wave of technology about to sweep our world. ~

Will Allen is an Inventor and Innovator at Hewlett-Packard. He is also an HP Fellow exploring the field of robotics and working on creating the workplace of the future.

The views expressed in this article do not represent the views of HP Labs, and they only represent the views of the author, Will Allen.

The New Wave of Robots

Jeff Burnstein

- President of the Association for Advancing Automation -

In a Silicon Valley pizza shop, a robot flattens a ball of dough into a perfect disc and passes it to another robot that spreads an ideal layer of sauce. A human worker adds toppings and passes the pie to a third robot that slides the pizzas into, and eventually out of, the oven. Not far away, a robot bustles through the hallways of a stylish, upscale hotel to deliver room service and other amenities to tech-friendly guests. And in a nearby shopping center, a robot barista named Gordon makes lattes and espressos for caffeine-starved locals who are too impatient to wait in line while human baristas chat with customers. In less-visible but more-common applications, robots vacuum floors, mow lawns, and clean windows in homes and offices across the country.

These are the next generation of robots.

Consumers and workers are familiar with traditional industrial robots on factory floors: fixed, warehouse-sized equipment operating at fast speeds behind safety cages that protect human workers from dangerous processes. But even this traditional industrial automation approach is changing with the advent of collaborative robots that are smaller, easier to program and use, and more flexible than traditional robots. While these so-called "cobots" can't replace large industrial robots for all applications, they are finding many new applications that traditional robots can't address effectively or cost-effectively.

Collaborative robots are designed with built-in force-limiting mechanisms so they can safely operate side by side with human workers, letting humans manage higher-level applications while the robots perform consistent, repetitive tasks. The robots' low upfront costs and fast return on investment make them attractive even for small manufacturers, who are increasingly competing on a global scale. Collaborative robots are now one of the fastest-growing segments of industrial robots. With that growth comes increased competition, which is helping to drive innovation, reduce costs, and increase acceptance. These changes are setting the stage for the next wave of robotics, which will occur closer to home.

The most common consumer service robots are vacuuming robots (think Roomba). Outside of those relatively simple and low-cost devices, few consumers interact with robots on a daily basis today, but that's already changing.

Retailers, hotels, and restaurants are recognizing the significant advances in productivity that automation offers, and they are looking for opportunities to automate mundane, repetitive, and undesirable tasks and move workers into higher-value roles.

For hotels that implement Savioke robots to deliver room service, for instance, the average 20 minutes that the robot spends navigating hallways and elevators to make a delivery is time that a human desk clerk can spend interacting with guests to improve their overall experience. Zume, the robotic pizza company, can make hundreds of pizzas an hour and has plans to expand beyond pizza to more widely automate the food industry. By automating low-value tasks, the company hopes to allow human workers to concentrate on activities such as recipe development and customer service.

The retail industry is also poised for dramatic change, and robot manufacturers are pursuing the opportunities they see there. Logistics and warehousing are already served by fleets of mobile robots, many of which are supporting the growth of online sales. But robots are also being explored by traditional retailers in their drive to remain relevant to consumers and competitive with online options. Retail robots range from basic kiosk-style mobile robots to SoftBank's Pepper robot — a humanoid robot that is programmed to chat and interact with customers, give directions, and answer questions.

Pepper — like Gordon, the robotic barista — has investors lining up behind these new human-interacting robots. To date, Savioke has garnered more than $17M in venture funding to support its goals, and robotic pizza company Zume has secured $48M. Pepper has received investments of $118M each from Alibaba and Foxconn, for a total of $236M for Softbank's robotics division. According to The Robot Report,[1] 2016 was a record year for investment in robotic startups, with $1.95B in funding. Service robots for professional and consumer use claimed two of the top ten categories for the number of companies funded.

While the use of service robots in retail, hospitality, and home applications like those described are still relatively new, they are expected to see significant growth over the next few years. The most recent report from the International Federation of Robotics (IFR)[2] expects that sales of service robots for professional use will grow an average of 20 to 25 percent from 2018 to 2020, for a total of about 400,000 units at a value of $18.8B. In the same time period, IFR estimates that sales of robots for domestic tasks could reach almost 32 million units, for an estimated value of about $11.7B.

Certainly, this new wave of robotics is in its nascency. Social acceptance may be an issue in some applications, but in general, consumers have responded positively to robots in a variety of hotel, retail, restaurant, and home applications. Long-term growth will depend on successful applications that can show strong return on investment and allow companies to optimize the technology to continue to evolve their businesses.

The year ahead — 2018 — will be a big year for robots coming out of the factories and into our everyday lives. ~

Jeff Burnstein is the President of the Association for Advancing Automation (A3), an umbrella trade group representing more than 1,000 global companies focused on robotics, machine vision, motion control, motors, and related technologies.

Figure: Automate Show in April 2017

Notes

1 https://www.therobotreport.com/2016-best-year-ever-for-funding-robotics-startup-companies/

2 https://www.therobotreport.com/22-research-reports-forecast-robotics-industry-growth/

The Rise of Service Robots

Steve Cousins
- CEO and Co-Founder of Savioke -

When it comes to predictions, roboticists can't help but dream big. Self-driving cars, in-home robotic servants, robocops, and other futuristic robot technologies are exciting, but they're still many years away from becoming widespread in our communities. However, one robot technology is already here and will gain even wider traction in 2018: service robots.

Autonomous service robots are those that improve the productivity and efficiency of their human counterparts. They're "co-bots" that work alongside humans to help them with specific tasks. They could be delivery robots, moving kiosks providing information or directions, roaming security robots, or any other type of robot that autonomously completes a task.

They differ from the industrial robots used for decades in manufacturing in that they roam about freely in busy public spaces such as malls, hotels, restaurants, hospitals, elder care facilities, offices, and high-rise apartment buildings. In comparison, most industrial robots are fixed on an assembly line, and the larger ones are kept behind cages to ensure safety. Building autonomous service robots introduces new technical challenges compared to their industrial counterparts, since they must constantly sense the environment around them to avoid collisions and operate safely in the presence of the general public.

So why are service robots here today and multiplying rapidly? Less than ten years ago, we entered a "perfect storm" in robotics. The cost to build autonomous robots hit a point that made them a viable commercial product. Low-cost sensors and chips, WiFi, advances in the open source Robot Operating System (ROS), 3-D cameras, and just-in-time manufacturing made it possible to build service robots quickly and relatively inexpensively. These new abilities vastly expanded the uses of robots in factories, but also created the opportunity to field delivery robots like the one my company manufactures — Relay — in hotels, apartments, logistics facilities, and offices.

Moore's law is also a key factor in the exponential growth in robotics. Named after Intel cofounder Gordon Moore, Moore's Law predicts that the cost of computing will be cut in half every 18 months.

Because of all these technological advances, in 2018, we'll see an explosion of service robots working alongside human beings, performing the dirty, dull, and/or dangerous jobs that people don't want to do. Industries such as hospitality, food service, medicine, and logistics are investing heavily in service robots to make their employees more productive and efficient. Our Relay robot will be working in hundreds of hotels, high-rise apartment buildings, and logistics facilities worldwide by the end of 2018.

We see huge demand for Relay as companies try to figure out how to get goods from point A to point B in indoor environments. All those meals and groceries delivered to apartment buildings by Uber Eats, Blue Apron, and Instacart? All those FedEx packages delivered to offices? All those high-value parts that need to move around a shop floor? Prescriptions in a hospital? These items are all now mostly delivered by humans, but that's a time-consuming and distracting practice for concierges, security staff, nurses, or other people who don't want to leave their posts to bring packages upstairs or across a building. That's why indoor delivery robots are taking off amazingly fast.

Land of Opportunity

Venture capitalists clearly see the opportunity for big profits with service robots. In 2016, they invested over $1.3 billion into robotics, with service robots being one of the fastest-growing sectors, and 2017 is likely to set an even larger funding record.

Of course, there are challenges to the rise of autonomous service robots. Hollywood has not done the industry any favors by creating "evil" robots that foment fear in people. The killer robots we see in movies seem determined to take over the world and make humans their servants. In reality, robots do discrete tasks that humans can't or don't want to do, and have no goals beyond completing those tasks. Artificial Intelligence (AI) has been extremely successful at creating systems that classify images from vast data sets (such as face detection or some types of object recognition), but even that field is a long way from creating a self-aware consciousness that could create its own nefarious goals.

Overcoming the human bias to fear robots is one of our biggest hurdles, but once people realize service robots are just "roaming computers," they'll accept them quickly, just as they accepted ATMs, laptops, and smartphones. Whenever we deploy a Relay robot, he almost immediately becomes "part of the team," and employees love working with him, treating him like a loveable mascot. Sometimes people just need to see robots in action to realize they aren't "scary" and are simply another technology that will change, but also improve, our lives.

Service robots don't mind doing tedious, repetitive, even dangerous tasks 24-7. And while they will help us be more efficient, service robots still completely lack compassion, negotiation, emotion, and the ability to know when to make exceptions. So there's little doubt humans will still need to be part of the equation, with robot/human teams leading to greater productivity, safety, and quality. ~

Steve Cousins is the CEO of Savioke. Steve hopes to measurably improve people's lives by creating sophisticated and friendly service robots that work safely and reliably in human environments.

Source: Savioke

Robotics and Automation in Construction Gain Momentum

Kaleb Steinhauer

- CEO of Genesis Dimensions -

Construction is one of the biggest industries in the world, and it is "likely to be one of the most dynamic industrial sectors in the next fifteen years" because it "is utterly crucial to the evolution of prosperous societies around the world."[1] These quotes by Fernando A. Gonzalez of CEMEX show how the construction industry is on the cusp of a revolution. The industry of automating construction is currently very small, but in the next year, I believe the construction industry will start to see increased investment in hardware technologies that focus on automating construction. This will lead to greater innovation that is concentrated on reducing the number one cost in the construction industry: labor.

If you think about it, the construction industry hasn't developed much since the beginning of mankind. Adam and Eve used a hammer and nails; today we are still using a hammer and nail.

A CEO of a very large construction company summarized it best when he said, "If someone would come to us today and say, 'We want you guys to build us a pyramid,' we would build it the exact same way as the Egyptians did 4,000 years ago — maybe add a crane to the process, but that's it." That kind of gives you the current state of construction. Also, over the past few years there has been little to no investment in innovation in the construction industry. There are only two VC firms that invest in construction technology, and much of that investment has been software focused. To achieve a significant gain in productivity, the construction industry needs new machines first; then it can focus on software that optimizes the new technology.

A lot of the initial technology around 3-D printing in automating construction was based on the gantry system in the image below. The gantry system is the more common version of a 3-D printer that most people have seen. It operates by using a frame around the area of construction that allows the robotic arm to print or construct objects inside the frame.

Source: Adobe Stock

People in the industry are realizing that these gantry systems are a thing of the past. They are too difficult to scale for large construction use and require intensive labor to set up and tear down before and after projects. Gantry systems have a lot of good applications, but they are just not practical for constructing something like a 1,000-foot-long warehouse.

Recently, companies have been starting to develop fixed robotics systems that are capable of 3-D printing for automating construction industries. My company, Genesis Dimensions, is one of only a few companies in the world that is working on fixed-access robotics systems that are capable of 3-D printing a building. There are a lot of companies that say they have the capability to do this and that, but they can't. The recent increase of investment activity in the industry has led to a lot of fast follower companies that have high claims for what they can accomplish, but do not have the actual technology under the hood. This increased investment will inevitably lead to some very influential technology in the automating construction industry.

The impacts of 3-D printing on the construction industry will be massive in the next 2-to-5 years. People will have greater access and choice in how they live. Labor is normally half of the cost of a construction project. The significant decrease in labor costs will allow for lower overall construction costs, leading to cheaper homes. People will also be able to design their homes however they want, and it's not going to necessarily cost more, because a robot doesn't care if it prints the same thing repeatedly, or if it prints something completely different each time. The robot just does what the software tells it to do.

These machines are also safer and could reduce the number of construction-related deaths.

This method of automating construction will also reduce the negative environmental impact that construction normally has. The number one contributor to landfills in construction, construction waste materials make up almost 30 percent of landfills. With an automated robot printing buildings, the amount of waste will be considerably reduced because it will only print as much material as is needed. The impacts of 3-D printing for automating construction will lead to more affordable, custom, and environmentally friendly development.

While the benefits of the technology are obvious, there will be some significant challenges to overcome before people just start 3-D printing buildings. The two biggest hurdles in the future will be dealing with government regulation and fighting a negative political perception. To get something like this approved may take some time, but it is important that regulators start thinking about it now. This is much like the debate over the regulation of autonomous vehicles that has started in recent years. Currently, the discussion on how to solve the labor shortage in construction is focused on, "How do we get more labor?" This is because most people don't even know this technology exists yet, so they need to be educated on the topic.

A lot of people's immediate pushback to 3-D printing in construction is that it will take away jobs. Unions are probably not going to like this technology, and some politicians won't like it either.

While regulation is a hurdle that must be overcome, it is critical that fighting the political battle to bring this technology to the mainstream be done in a way that highlights the positive impacts it would have.

There is an incredible amount of opportunity to leverage new technologies to enact positive impacts in the construction industry. It's almost like shooting at a dartboard in the dark, and you hit the bullseye every time, because whatever you're going to do in this field or sector, you've got a good shot at being successful. The construction industry is still very far behind the rest of the world, and it is just beginning to look at how increasing investment in innovation can help close the gap.

What the industrial revolution did for manufacturing is akin to what robots and automation could do for construction in the future. And in 2018, the level of investment, interest, innovation, and progress in construction robotics is likely to accelerate significantly. ~

Note
1.https://www.usnews.com/news/articles/2015/11/10/america-to-help-lead-global-construction-renaissance

Kaleb Steinhauer is the CEO of Genesis Dimensions. Kaleb's solution combines large-scale 3-D printing and robotics to revolutionize the construction industry through lower labor costs and limited environmental impact.

Hyperloop and The Future of Transportation

Holly McNamara

- Executive Chairman of Hyperloop Massachusetts -

An individual's biggest asset is time. Everyone is always looking for ways to have more time in their day. In 2018, new technologies focusing on automation in the transportation industry will become increasingly mainstream as they move from proof of concept to minimum viable product.

Transportation is one of the industries that is the most ripe for disruption due to robots and automation. In 2016, a report from the Executive Office of the President titled *Artificial Intelligence, Automation, and the Economy* said that almost 3.1 million jobs in transportation could be threatened or substantially altered by technology. In the next year, these technologies are only going to become more and more present in everyone's lives.

In 2017, there was a lot of progress in the transportation industry that may lead to the beginning of a transportation revolution. Some technologies proved proof of concept, and some even achieved minimum viable product. The Hyperloop, which was initially proposed in a white paper by Elon Musk in 2013, is a futuristic method of transportation that uses magnetic levitation to propel pods through airless tubes. In April 2017, Hyperloop One completed its first successful test run, establishing proof of concept for the first time.

Hyperloop One's momentum continued when it announced a partnership with Virgin Group, a Richard Branson company. This partnership showed serious investment is being made in this technology, and several countries around the world have indicated their interest in the Hyperloop as well.

Automated vehicle technology, more commonly referred to as self-driving technology, showed that it has officially reached minimum viable product or maybe even a little more than that. Semi-autonomous driving is a feature available in some newer cars on the market today. This technology works using sensors with Lidar (light radar) to constantly monitor the environment around the car as advanced software processes all the inputs and plots a path for the car. This technology is already a hotly debated issue among regulators in government at a national and local level, because companies are eager to get their technology out first. Arguably the most important technology-focused legal battle of the decade is playing out currently as Uber, Waymo, and Alphabet Inc. are currently engaged in litigation concerning trade secrets.

Autonomous vehicle and Hyperloop technologies are just two of the major technologies coming soon. Looking further ahead, one can even envision more fantastical-type technologies such as consumer space travel and off-world colonization.

In 2018, people will start to see new forms of transportation technology become increasingly part of the norm in their everyday lives. The need for on-demand transportation will rise, with more people using ride-sharing services. It appears that Uber is preparing for a transition to a partly autonomous fleet. The company has a small-scale fake city in which it is rapidly testing self-driving vehicles — along with several pilot programs in cities across the United States. If regulations were not an issue, these might already be on the road today. Dubai has already tested autonomous flying drone taxis. These new technologies could have an immediate impact on people and businesses. Autonomous vehicles will make transportation easier and potentially safer. People will have a greater freedom of choice. Corporations will be able to transport goods much more efficiently, and small businesses will even be able to take advantage of automated distribution networks. In the next year, the Hyperloop will start to gain steam as countries compete to build the first successful Hyperloop network. This will be driven by both the impact it would have as a passenger transport system and as a potentially significant disruptor of long-haul transport.

Before these technologies can be more widely adopted, however, there are several risks and challenges that must be overcome.

There are safety and security risks that accompany any new method of transportation. Autonomous vehicles should go through strenuous testing in order to prove they are ready for mainstream use. While estimates show autonomous vehicles to be a safer mode of transportation, the big question is, what will a mix of human drivers and autonomous vehicles look like? The biggest benefit from these vehicles will be realized through their "connectedness" or their ability to communicate constantly with the internet. However, this technology could also be hacked, posing a significant security risk. The first challenge to overcome will be regulation. To get the technology in front of the mainstream consumer, companies need to get the government on board. Previous years have shown some bitter battles between ride-sharing companies and local governments. Uber and Lyft have pulled out of cities due to regulations. Transportation technology will have to prove that it is safe and reliable before regulators will allow it to become mainstream.

While autonomous vehicle technology is almost here, ideas like the Hyperloop and other more fantastical technologies are a little further away. Most new transportation technologies will face similar risks and challenges, but they hold the potential to have a considerable disruptive impact on many aspects of society.

In 2018, the "idea" phase of the transportation revolution will start to fade, and the "material" phase will begin its rise. ~

Holly McNamara is the Executive Chairman of Hyperloop Massachusetts. Holly is working to bring revolutionary new forms of transportation to the U.S. East Coast and beyond.

Image source: Hyperloop One

Robots in Hospitals and Pharmacies

Stanislaw Radominski

- Vice President of UnitDoseOne -

The automation of various often dreadful and highly repetitive tasks is a fact in many industries. The total number of professional service robots sold in 2016 worldwide rose considerably by 24%. Medical robots, as well as pharmaceutical systems and logistics robots, which belong to this group, have the largest share and the strongest growth dynamics in the number of units sold (34% increase in 2016).

Sales of medical robots increased by 23% in 2016, accounting for a share of 2.7% of the total unit sales of professional service robots (World Robotics 2016 Service Robots). Surprisingly enough, this is not really the case when you look at drug-preparation automation (also referred to as pharmacy automation), especially in hospitals.

The growth in hospital pharmacy automation is rather stagnant. Only about 8% of U.S. hospitals used robotic solutions in 2014, and this number has not really improved since.

Pharmacy robots can be divided into two groups: inpatient preparing and dispensing patient-specific drugs in single doses, and outpatient/retail systems that fill and dispense the prescribed drugs. A separate category exists for home solutions.

Inpatient automation seems to be stagnant, and most probably not much is going to change in 2018. Right now, almost all U.S. hospitals use automated dispensing cabinets (ADCs) in their medication-distribution systems (ASHP, 2014). They are, however, not really that automated, usually employing "pick-by-light" systems with manual drug retrieval. Having many advantages over the centralized systems (drug is on demand right next to the patient), the decentralized drug dispensing has not seen real automation so far. The doctor still has to prescribe the medications for the patient, those have to be packed into an ADC tray by a pharmacy technician, and the tray has to be delivered to the machine at the ward and then manually picked up by the nurse at the ADC unit and finally delivered to the patient's bedside. Only a fraction of hospitals employ robotics unit-dose centralized solutions. The rising demand for accurate and timely dispensing and the elimination of dispensing errors will be the driver for a change in this sector.

2018 might be the year where some new and interesting solutions will try to win the stage.

Expiration of several patents on storage and dispensing systems and mechanisms, together with expansion of robotic devices, should lead to new constructions designed for smaller and medium hospital pharmacies. A few U.S. patents on the automated system for selecting and delivering packages, like those owned by Automated Healthcare (currently a division of McKesson), expired recently, releasing to the public domain some ideas based on unit dose. Moreover, robotic technology is gaining popularity, accessibility, and constant miniaturization. Vision systems are more and more powerful and cheaper, motors smaller and more powerful, and controllers faster. These opportunities can lead to the construction of modular robotized pharmacies that are feasible for medium and small hospitals.

One example of a technology that has the chance to gain ground is a modular and scalable robotic pharmacy system, which can be tailored to a specific hospital's needs, offering real dispensing automation. This is what we have been working on, and we believe there could be more adoption this year. One of the things to back this up is the fact that one of the health systems in Ohio wants to pilot our solution in early 2018 to verify its potential and later expand it to all of their hospitals. The idea of having several ward units with a high level of automation saves time from faster dispensing (most of the prescribed patient sets can be prepared in advance, instead of preparing them all on demand) and resources (the money saved on tightened drug-dispensing security or different deployment of staff currently doing repetitive and error-prone tasks, like pill counting or tray filling).

The idea is based on combining automation in several stages at a high level of sophistication but enclosing the physical interface unit in the actual machine — into a size that is similar to a current ADC unit.

Aside from what we have been working on, we have also seen a push toward automated units that combine packages of on-demand, patient-specific oral-solid medications with the ADC units that stock unit-dose medications and larger items (like alixaRx) .

One important signal for the higher adoption of automation in the pharmaceutical robot segment is a recent push by the World Health Organization, with the launch of its third global initiative, Global Patient Safety Challenge on Medication Safety. It aims at halving the avoidable medication-associated harm by 2022. This means that behind the push to save money and to find more efficient ways to dispense drugs, there is also a growing mandate to improve medication safety. This is a big deal, and more people counting pills isn't the solution. Robotics is the solution — and there isn't just one solution here. A lot of companies are working on this problem.

This initiative is going to drive investment and advancements, and 2018 is a year when we could see traction for automated medication solutions increase. By 2022, of course, it's likely to be moving fast because of this WHO mandate. But 2022 isn't that far in the future, so 2018 is going to be a big year for ramping up — and for increased awareness about medication safety.

For outpatient and retail experience, 2018 is likely to show an increase in robotic systems adoption. Ever-growing competition between retail pharmacies and an increasing workload for pharmacists are the motors of change here. Prescription-filling machines and automated storage units are becoming standard in many facilities. And we've already seen this advancing with mail-order drugs. Several big players are already present on the market, with PillPack (which uses oral-solid packing machines to prepare monthly prescriptions in ready-to-administer packs) being recently named among *Forbes*'s list of "The Next Billion-Dollar Startups" in 2017. Clearly there is a lot of opportunity to automate this space, and there is a lot of societal and financial value that can be created in the process.

Having robotic and automated prescription filling available in ready-to-administer systems is going to be an increasingly hot topic for big pharmacy chains. And it could begin to get more attention in 2018.

Such systems can ensure medication adherence with regard to dosage and medication, but they still cannot ensure compliance. This is the third trend in pharmacy automation, which has been steadily developing in the past few years. There have been many attempts to fix the compliance issue (mainly though mobile applications), but none of these seems to have been all that successful. This is where there is an opportunity for home drug-dispensing robots, like Pillo, which help advise the user on how to take their medications.

These technologies will develop over time, but this opportunity may take longer to see adoption than the hospital-based, online, and brick-and-mortar solutions that are focused on prescription adherence. As more robots enter homes, however, proper drug dispensing could become an important service. ~

Stanislaw Radominski is the Co-Founder of UnitDoseOne. He is disrupting the drug distribution process by using advances in robotics to automate pharmacies and hospitals around the world.

Source: The Futurist Institute

Robots and Automation on the Road

Craig Fuller

- CEO of TransRisk -

2018 will see an accelerated convergence of automated and autonomous technologies in the trucking and freight industries. So far, many promising systems have been developed separately, in a piecemeal way. But next year, the trucking industry will begin experiencing transformative changes as these various advances are integrated with one another.

Start-ups and established enterprises have developed self-driving and platooning systems that will start bringing measurable gains in safety and productivity in the near future. Electronic logging devices — mandated for every truck over 10,000 pounds — will allow greater visibility into the movement of carrier assets and shipments, and the most robust of these devices give further insight into the performance of specific truck components like brakes, drivetrains, and oil and tire pressure. And they will be able to send alerts directly to the maintenance garage.

Other tracking devices are being implemented in trucking that attach directly to packages, monitor their location and condition, and ping remote operations centers on an hourly basis. Automated port terminals like those already in place in Los Angeles and Amsterdam will become more widespread.

Technologies that enable transparency and automation will proliferate throughout the industry. Customers will demand more information about their freight's location, for the purposes of security, customer satisfaction, and liquid supply chains. With higher interest rates on the horizon and pressures on costs coming from labor and commodity prices, organizations will be under pressure to manage these costs. Customers spoiled by the "Amazon effect" of faster delivery cycles will demand quicker turn-arounds and on-demand offerings.

In years past, much of the customer experience was related to UI/UX, and shipping was relegated to an afterthought. Amazon proved this to be a mistake. They used their supply-chain and logistics operations to gain a substantial edge on the competition, providing customers with near-time gratification that previously was only available through brick-and-mortar.

Blockchain proliferation in the industry is also expected to happen in 2018. A group of hundreds of top logistics players have come together to form a blockchain consortium known as the Blockchain in Transport Alliance (BiTA).

The BiTA organization will develop industry-wide standards for blockchain-enabled smart contracts that create trust between unknown actors, increase visibility into the movement of trucks and goods, eliminate fraud, and streamline payments.

Today, asset-based trucking carriers devote enormous resources to the management of drivers: large fleets often contend with 100% annual turnover in their driver pool and have to offer generous signing bonuses and reimbursements for training programs to attract new drivers. Trucking is a $700B industry, and a third of the costs go to compensating drivers.

As trucking becomes more automated, these carriers will shift toward data-intensive analytics to add value to the supply chain. Not only will wayfinding — to avoid congestion and find cheap fuel prices — be automatically optimized, but even trucking contracts themselves will be sorted, negotiated, and closed with the assistance of algorithms manipulating data sets that are only now being assembled and collated for the first time. Voice brokers, freight forwarders, and 3PLs with weak IT infrastructure that lack transparency will see themselves disintermediated and replaced by automated digital solutions.

Also in 2018, a new futures market is being created for trucking spot rates. These cash-settled futures contracts will enable trucking market participants to hedge their shipping rates. This will, in turn, help to create index-linked contracts and markets, further enabling pricing and capacity transparency.

Smart trucks will communicate with each other, infrastructure, manufacturers, service providers, and other entities through the Internet of Things (IoT). As advanced telematics become ubiquitous in "smart trucks" and parts are continuously monitored and replaced before failure, trucks will start to resemble services more than products — carriers will pay for digital solutions to keep their trucks in tip-top shape and moving down the road rather than taking on unpredictable and wasteful maintenance costs themselves. ~

Craig Fuller is the CEO of TransRisk, where he is creating a futures and options market for transportation spot capacity and working toward providing industry participants with a transformative new way to protect against price volatility on major U.S. trucking routes.

ChatBots That Talk

Lorenzo Carver

- Founder of SpeadBot and CoinSwipe -

Conversation robots, or chatbots, are computer programs that conduct a conversation or perform a task via text. Studies show that people who interact with a conversational robot tend to expect the robot to do human things as well as be able to say human things. Chatbots can build smart contracts using blockchain technology through something as simple as a text message conversation. Companies and consumers can use this technology in a variety of ways, from conducting business to cryptocurrency investing. In 2018, conversational robotics will move forward to mass adoption with the introduction of the first 'killer' chatbot that uses blockchain technology to make a task easier and faster for the user.

Despite the hype and press you see around conversational robots, most of the people using chatbots *consciously* today are developers.

The clear majority of people who use mobile devices don't consciously use chatbots, but they may be interacting with them unknowingly. One of the biggest areas we currently see chatbots being used in is customer service. Many of the chat boxes that pop up on the screen offering to help a customer when on a retail website are not humans but conversational robots. The initial information is taken by machine, and based on how the conversation progresses, a human may step in. So basically, the chatbots help guide the consumer to a funnel, and depending on how the interaction proceeds, it can either route the consumer to an appropriate representative to try to turn the lead into revenue for the company, or if revenue generation is unlikely, the company may just allow the consumer to continue talking to the robot.

This process is similar to how a call center with humans operates. When an inbound lead is generated, the first thing a system (including a human representative) does is gather information about the consumer to see how they can get the potential customer to the resource most likely to optimize the outcome for a company. While conversational robots are currently used for sales and customer support, the future will bring chatbots that make other tasks faster, easier, and more cost efficient.

When chatbots were first coming about, a lot of the interest and focus was around the idea that these robots would be able to understand natural language and have a fluent conversation with people. While technology has advanced rapidly bringing us very close to this, doing it in a general-purpose way that might apply to a wide range of scenarios has been problematic.

Companies would claim to have conversational robots that were artificially intelligent, but very quickly it became clear that their bots didn't have the capability to understand the nuances of human conversation. The technology has begun to transition to take out conversational aspects and just turn these chats into a general-purpose user interface that looks very much like texting or using a messaging platform. Now that companies have realized this is the best way for consumers to interact with conversational robots, they will begin to develop the 'killer' bot that will lead to mass adoption — much like SMS did for smartphones or Excel for the PC.

In 2018, people are going to see conversational robots married with blockchain technology. Billions of people who have internet access don't have any cryptocurrency, don't know what cryptocurrencies are, and they don't know what blockchain is. This is literally changing the foundations of how commerce is conducted in the world. And it will become more apparent to the average person in the year ahead. ~

Lorenzo Carver is an Inventor and Entrepreneur currently developing conversational robots. He is working to combine chatbots and blockchain technology to make complicated takes easier for consumers.

Investing in Robots

Jeremie Capron

- Director of Research at ROBO Global -

For investors in Robotics, Automation, and Artificial intelligence (RAAI), it's an exciting time. In the first 11 months of 2017, the ROBO Global Robotics & Automation Index — the leading stock index created in 2013 to track the robotics and automation revolution for investors — returned an incredible 47%, significantly outperforming broad equity market indices. The financial markets have made it clear that RAAI may be one of the most important investment opportunities of our generation.

Perhaps even more impressive is that this multi-decade technological transition is still in its infancy — and the ingredients for a major breakthrough are already in place. Costs are rapidly declining across key enabling technologies such as computing, sensing, and communication. At the same time, we are seeing an explosion in terms of performance capabilities that dramatically expand the scope of RAAI and how it is applied in nearly every industry.

Perhaps most importantly, the world is generating an almost unimaginable quantity of the fuel that powers AI: Big Data. And that pace is growing at a frenetic pace of *billions* of gigabytes every day.

Automation itself is nothing new. We have been automating the "dull, dirty, and dangerous" work of humans for decades, from the earliest agricultural machinery to today's high-speed welding robots used in manufacturing. What's changed is the advent of machine intelligence that has unleashed a massive and exponential wave of change. As a result, RAAI technologies are shifting the fundamental structure of every sector of the economy, altering the way we grow and process our food, how we manufacture and move goods of every kind, how we shop and play, and even how we learn and care for our health.

This revolution is happening now, and it promises to be even more transformative to our society and how we function than personal computers, the Internet, mobile devices, and big data before it. At ROBO Global, we are not alone in our belief that RAAI's economic impact will be measured in the trillions of dollars. Andrew Ng, then Chief Scientist of Baidu and Adjunct Professor at Stanford who led the development of its Massive Open Online Course platform, has called Artificial Intelligence "the new electricity." Acclaimed futurist Ray Kurzweil has said that "Artificial intelligence will reach human levels by around 2029" and that by 2045 "we will have multiplied the intelligence, the human biological machine intelligence of our civilization a billion-fold." That's quite a revolution.

Clearly, corporate giants around the world understand this vision of the future and are aggressively deploying capital to develop and acquire the RAAI technologies they need to take their businesses to the next level. The choice is to be a disrupter — or to be disrupted. As companies rush to leverage and build on all that RAAI has to offer, the pace of RAAI acquisitions has continued to accelerate. In November 2017, Emerson Electric unveiled a third bid to acquire Rockwell Automation, the largest pure-play industrial automation company outside Japan, for an astounding $29B, more than double its market value at the start of 2016. (Despite what looked like a very rich valuation, it was not enough to convince Rockwell's board to accept the offer.)

Earlier in 2017, Intel paid $15B to acquire the driverless car technology company MobilEye. These are just two examples of the dozens of transactions that were initiated in 2017 — all on the heels of 2016, a banner year in which more than 50 robotics and automation companies were acquired for a total of more than $20B. Note that nearly half of the 2016 acquisitions involved Chinese money, including the $5B acquisition of German industrial robotics company KUKA by China's leading appliance maker MIDEA. And that total does not include key software plays in industrial automation, such as Siemens' $4.5B acquisition of Mentor Graphics, and companies like Google, IBM, Yahoo, Intel, Apple, Samsung, and Salesforce, who are all competing head-to-head in an acquisitions race, with more than 40 deals announced last year alone. For investors, the name of the game is to future-proof their portfolios. And while venture capitalists have clearly shifted their focus toward RAAI, it looks like public equity investors may need to play catch-up, especially in the US.

On the private, earlier stage company side, robotics startups had a record year in 2016, with deals up by nearly 20% to reach more than 170[1], and totaling $3B since 2012, excluding pure software companies and driverless cars. In 2016, over 550 startups using AI as a core part of their products raised $5B in funding.

In public equity markets, investors have only just begun to approach RAAI as an investible theme. Since the 2013 establishment of the ROBO Global Robotics & Automation Index and the flagship ETF that tracks the Index, a handful of competing thematic strategies have come to market around the world, totaling, by our own estimates, approximately $23B in assets under management. This is just a tiny drop in the bucket of the global equity markets, which total more than $76T. It also looks small in the context of private initiatives like Softbank's $93B Vision Fund — the largest tech fund in history.

Following the money by region reveals very interesting dynamics. Indeed, US investors seem to be lagging considerably behind European and Japanese investors when it comes to flowing money into RAAI. Our analysis suggests that Japanese and European investors have allocated more than $10B each to dedicated funds. In contrast, it appears that US investors have allocated just $3B into funds that are focused on RAAI.

It is important to note that not all RAAI funds are created equally. While each may offer investors the opportunity to direct funds toward the RAAI revolution, it's important to recognize that these funds pursue vastly different investment strategies, and some are significantly riskier than others.

Certain funds invest in just a handful of large-cap stocks in the sector in the hopes that these will be tomorrow's winners, yet it's very possible — if not highly likely — that in today's dynamic environment, the largest players will change nearly as quickly as the technologies themselves. For that reason, a strategy that includes broad exposure to the global value chain may be a much more prudent approach.

This broad exposure may include not only large-cap participants but also providers of key enabling technologies such as sensing, computing, and actuation, as well as providers of applications that deliver capabilities in every industry, including factory automation, surgical robotics, food and agriculture, 3-D printing, logistics automation, and more.

In the words of Danish physicist Niels Bohr, "It is difficult to predict — especially the future." This is especially true in the world of robotics, automation, and artificial intelligence. Never before have we seen such an explosive mix of exponential technological change, an abundance of disruption, and high levels of mergers and acquisitions that are driving massive moves in share prices. Who could have forecast that the 3-D printing sector would appreciate a stunning 8-fold from 2012 to 2013 — or foresee its 80% collapse in the following two years? Yet shifts of this magnitude are inevitable as RAAI technologies continue to evolve at an unprecedented pace. Investors who grasp this reality will choose to seek superior risk-adjusted returns by building a portfolio of industry leaders that is diversified across geographies, small and large companies, and technologies and applications.

As well, investors who understand that selecting these industry leaders requires careful, informed research is crucial will rely on the guidance of the industry experts, academics, and entrepreneurs that can provide the knowledge and insight to more accurately foresee emerging trends, identify the most promising new technologies, and understand the intricate interaction between technologies and their specific applications.

Across industries and around the globe, companies are revising and rethinking their own strategies to cement their futures in a world that is dictated by robotics, automation, and artificial intelligence. The financial markets have already recognized this shift and have begun to reward those who are placing their bets on the future. For investors who are seeking a strategy to capture those rewards — and truly future-proof their portfolios — the time is now to invest in all that RAAI has to offer. ~

Note

1. CB Insights

Jeremie Capron is Managing Partner and the Director of Research at ROBO Global. ROBO Global helps investors capture the growth and return opportunities presented by the megatrend of robotics and automation across industries.

Robots in Energy

John Gibson

- Head of Disruption at Tudor, Pickering, Holt & Co. -

What exactly is a "robot"? For the sake of market hype, the term is often so broadly defined that almost any automated or electronic device seems to qualify. Who can distinguish between a robot and a "machine," an intelligent machine, an "appliance," or an intelligent appliance? Perhaps that task is best left to a taxonomist, or worse, an ontologist. Instead of coining some cool new word to sell yet another piece of arcane machinery, let's settle for a more straightforward definition. A robot is simply an electromechanical device directed either by electronic circuitry or by a computer.

Based on this definition, the energy industry has been using robots for decades. Why? The value proposition they offer the oil and gas sector falls into three broad categories: safety, quality, and cost.

Deploying the next generation of robots will dramatically increase value in each of these areas.

While robots will eventually impact every aspect of oil and gas operations, consider the following four examples: (1) the reinvention of an old staple: welding and machining, (2) a newcomer inspired by space exploration: autonomous underwater robotic vehicles (AUVs), (3) a new approach to offshore seismic acquisition: recording autonomous vehicles (RAVs), and (4) a safer, faster, more reliable rig floor pipe-handling system: The Iron Roughneck.

Welding and Machining
Hundreds of small welding and cutting boutiques exist to meet the demands of the oil and gas industry. Each shop varies in terms of expertise and equipment. Some focus on custom work, producing a single replacement part that may be unavailable for extended periods of time.

Custom welding and machining attracts the best craftsmen in the industry because it usually involves low volume, highly profitable work completed under a tight deadline. What's more, it requires an expert welder or machinist to create a specialized one-of-a-kind part. Perhaps one day, 3-D printing will invade this custom space. Today, however, constructing and verifying a new custom part in a CAD/CAM system takes longer than simply having a craftsman/artist build it.

On the other hand, some shops focus primarily on contracts to manufacture large quantities of a particular component.

The best craftsmen shy away from mass manufacturing, because the work tends to be rote, boring, and riddled with quality issues no matter how intensely managed the process may be. Couple the repetitive nature of the project with a tired, stressed, or impaired employee, and less than adequate components are often produced, diminishing profitability and/or compromising safety.

These are good reasons to replace at least some people with robots. In fact, the welding industry has been steadily improving robotics for more than three decades, striving to eliminate risk factors associated with rote jobs.

Boredom and inattention are not the only aspects of mass machining that increase risk. Often, a large flare tip, boiler, blowout preventer or another piece of equipment requires several shifts to complete construction. This creates the opportunity for a single component to exhibit significant variations in welding quality, depending on which workers were involved. Robotic welding eliminates inevitable variations that numerous people introduce.

The ultimate goal is to guarantee "x-ray" quality welds connecting every piece. Without investing in costly nondestructive examination of every weld, the only way to achieve this is by deploying properly configured robots. Robotic welding eliminates human variations in emotion, fatigue, impairment, distraction, health, the need for a vacation, and changes of shift.

The ongoing evolution of CAD/CAM systems will inevitably enable even rapid production of custom parts. Not only are these systems becoming easier to use, but "infinite" storage means the necessary files for any new part will be available in perpetuity, making retrieval for manufacture a simple task.

People will not become obsolete, however. In the future, great welding craftsmen will remain essential to the maintenance and longevity of the industry's legacy assets.

Autonomous Underwater Robotic Vehicles

Robots offer even greater promise in costly offshore operations. Research into autonomous underwater vehicles (AUVs) and autonomous surface vehicles (ASVs) began as early as 1957 at the University of Washington. Today, a vast number of new unmanned technologies continue to emerge.

Every year, the offshore oil and gas industry spends $2 billion on remotely operated vehicles, all of which require ships, personnel, and long cumbersome umbilicals. In the Gulf of Mexico alone, over 43,000 miles of active and inactive pipelines lie on the bottom of the gulf. Approximately 53,000 wells also need regular inspection and maintenance. The North Sea and other global offshore operations substantially increase those numbers.

With today's lower oil and gas prices, offshore oil and gas markets delight in new technology designs that promise to reduce operational costs, enhance safety, and improve efficiency.

Into this huge existing market, Houston Mechatronics plans to deploy innovative new solutions that will literally cut the cord from today's manned ships and remote controls. Designed by former leaders from NASA, based on designs inspired by robots and remote communications for space exploration, a new generation of autonomous underwater robotic vehicles (AUVRs) will help launch the offshore operational market into a new era.

This new generation of autonomous robots will more efficiently and cost-effectively resolve pipeline integrity issues in state and federal waters. Imagine "robots as a service" working almost continuously to inspect and repair multitudes of subsurface facilities for both operators and governmental agencies conducting audits.

Recording Autonomous Vehicles

Simultaneously, Kietta is constructing a new generation of offshore seismic acquisition vessels. These recording autonomous vehicles (RAVs) deploy robotically controlled recording cables below wave depth, decoupled from the seismic source. This new capability has the potential to reduce noise and increase data density in areas where operators require higher resolution. The reduction in noise associated with cable flow also could enhance data quality. The most significant upside may lie in making 4-D acquisition offshore much more affordable.

The Iron Roughneck

For decades, the oil and gas industry has been on a trajectory to eliminate near misses, incidents, and fatalities in one of its most active and hazardous work areas: the rig floor. NOV has been a leader in pipe-handling innovation since 1975, when it introduced "Bigfoot," the first generation of "Iron Roughneck."

Creating robots capable of making and breaking (screwing and unscrewing) pipe connections has eliminated the dangerous technique of "spinning chain" to tighten and loosen pipe connections. This probably caused more hand injuries than any other manual operation in the history of drilling. Today, a remotely operated robot controls the spinning and torque of tubulars, eliminating the entire process responsible for so many severed digits in the past.

In long horizontal wells today, torques involved in making and breaking pipe connections exceed 75,000 foot-pounds. Robots continue to become both more powerful and more compact, enabling easy retrofit of existing rigs without having to increase the size of the drill floor.

Current investments in the robotization of rigs foreshadow an era in which "self-drilling" rigs may proliferate even before auto manufacturers overcome the issues impeding the adoption of self-driving cars.

Conclusion

The energy industry has long been an eager consumer of robotic ingenuity. As we look to the future, operators, service companies, and technology entrepreneurs must continue to focus on sustainability. Robots will provide continuous breakthroughs in every aspect of oil and gas operations from the rig floor to the seafloor. ~

John Gibson is a Senior Advisor focused on Disruptive Energy Technology at Tudor, Pickering, Holt & Co. John is working at the forefront of energy technology, bringing together executives from corporations to push the industry forward.

Robots in Finance

Jason Schenker

- Chairman of The Futurist Institute -

Automation has been a growing part of finance in recent years, and its presence is likely to grow in 2018. Asset management has long been dominated by computers, statistical analysis, and programming. And financial technology — known in shorthand as FinTech — has been disrupting asset management with passive trading strategies. Some of these strategies are known as Roboadvising, due to their automated (i.e. robot-like) nature. And the disruption potential for asset management is very high.

These aren't the kinds of physical robots you might see in 2018, but Roboadvising is likely to grow in the year ahead. In fact, FinTech solutions are likely to spread throughout financial services. For the average person, FinTech solutions like crowdfunding, digital currencies, payments, and Roboadvising generally reduce costs, reduce complexity, and increase ease of use for transactions that had previously been the domain of banks. Many of these offerings became more visible in 2017.

Roboadvising has the potential to be particularly disruptive. A survey conducted by the CFA Institute highlighted the disruptive potential from financial advice tools, and the impact on asset management was at the top of the list of sectors likely to be most affected. This disruption is likely to accelerate in 2018.

Sectors Most Affected by Financial Advice Tools

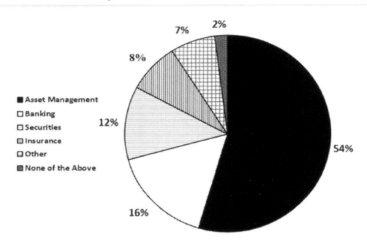

Legend:
- Asset Management
- Banking
- Securities
- Insurance
- Other
- None of the Above

2%
7%
8%
12%
54%
16%

Source: CFA Institute, Prestige Economics

FI THE FUTURIST INSTITUTE

Passive asset management techniques and Roboadvising are often easier and cheaper to administer than active asset management. These strategies can be implemented at significantly lower costs than active asset management strategies, because they no longer require human asset managers. There is an economy of scale, when computer programs do all the strategy work, analysis, and planning, as well as all of the buying and selling of securities.

Passive asset management has also been adopted by finance and trading, because these fields have historically embraced technology, with many firms using black box, algorithmic, and technical trading strategies for many years. Expensive items (like market research) are also no longer parts of the budget, since decisions are made by computers. After all, trading computer programs do not read words. But they really like lines — especially lines above which (or below which) the price of a traded security has consistently stayed for a long time.

In the figure below, you can see gold prices with some critical trading technicals. The image shows how gold prices fell hard and fast after closing just below an upward-sloping technical support (upward-sloping blue diagonal line) that had been in place from December 2015 until the end of September 2016.

Gold Prices Show the Importance of Technicals

Source: Prestige Economics, eSignal

PRESTIGE ECONOMICS

Prices were above that line for a relatively long time. And it was a line the computers — the trading robots — were watching. It was also a level that I highlighted in our research frequently over the course of many months. As you can see in the chart, before a big selloff, there was a price drop below the upward-sloping diagonal line.

Essentially, technical traders respected this diagonal line, and once it was broken, gold began selling hard and fast. This kind of trading drives roboadvising, and is likely to become an increasing part of financial trading in 2018 — and in years to come.

Technical trading has become more important, so analysts have been trying to add value by knowing what lines and supports matter most for the computers in different markets. This is why there has been a significant increase in the number of financial professionals pursuing the Chartered Market Technician® designation. The CMT® — a designation I completed in 2016 — focuses exclusively on these kinds of technical trading dynamics. Essentially, you are looking for the computers in the market.

I expect these kinds of trading dynamics will become increasingly important in 2018 and beyond, as passive asset management and Roboadvising continue to expand. But there is one big risk to Roboadvising: it has not yet been tested by a market downturn. Even though these are largely automated solutions, they are not all the same. And some of them allow for people to interfere with their trading algorithms.

It remains to be seen if people will let the robots do all the trading when a market is in freefall, or if they will lose their faith in trading automation. When money is on the line, people's behavior can change. And as Michael Walton has noted in his essay in this book, people still don't fully trust robots and automation. ~

Jason Schenker is the Chairman of The Futurist Institute and the Editor of *The Robot and Automation Almanac*.

This is an excerpt from the book *Jobs for Robots: Between Robocalypse and Robotopia*.

Building the Right Robots

Bruce Welty

- Chairman of Locus Robotics, Inventor, Innovator -

The fulfillment industry is one of the most underappreciated industries in the world. In today's world, people want to get the things they want delivered to their doorstep. The movement from bricks to clicks is a zero-sum game — whatever business retail stores lose results in a dollar-for-dollar gain in e-commerce.

The rise in e-commerce is only just beginning, and the augmented delivery of goods will start to affect people's everyday lives. The fulfillment industry is what makes this possible and is possibly one of the most underappreciated industries in the world. This industry is responsible for the process of receiving, packaging, and shipping orders for goods. So, anything consumers order online and get delivered has some form of fulfillment to go through. It could come from a very large warehouse distribution center or just from the back room of a small boutique selling products online.

A lot of companies with large fulfillment needs have already started to implement robots and automation to increase efficiency and decrease costs. In 2018, the fulfillment industry will see forward leaps and bounds in the optimization of the already existing technology driven by society's increased desire to get the products they want delivered to their door.

The watershed moment of the past year was finally knowing and understanding the X-Y coordinates of everything in a warehouse or other fulfillment facility. Previously, a company could set up a warehouse and optimize it, but as time passed and warehouse racking systems were shifted around, the integrity of the original programmed coordinates would downgrade. Now, the efficiency can be maintained and even improved because companies can track the X-Y coordinates as they change and optimize continuously around that. One significant driver of this is increased accessibility of the sensor technology needed. The rise of technologies like automated vehicles has led to increased development in the sensor. This has resulted in many new types of sensors being developed and offered at a much lower cost. The newer Lidar (light radar) sensors provide a higher resolution of the physical world than previously available, and with updated software available to analyze it, companies can optimize warehouses to a greater degree than ever before. This allows warehouses to discover better ways to group and sort orders that can help to streamline the distribution process. Some of the changes Locus Robotics has implemented result in 20% to 30% improvements with one small tweak to an algorithm. This type of analysis is only scratching the surface of the optimization that could be possible.

While some parts of the industry have been quick to embrace new robotics and automation strategies, others have been staunchly resistant. One of the biggest hurdles this type of technology faces is skepticism. When people don't completely understand how the technology works, or it confuses them, they immediately reject it, and in the warehousing industry, people are accustomed to serial processing. They don't see that 100 robots moving at 1 mile per hour is the same as 1 conveyor belt moving at 100 miles per hour. The industry must get used to the idea that even though the fulfillment doesn't look as frantic or fast as serial processing, it's extremely fast and very calm. Another hurdle to overcome will be regulatory challenges. Could future taxation changes affect this type of labor? Are safety regulations going to change? The industry will need to embrace the technology to keep up with the demand in the future, and companies that refuse to adapt will get left behind.

The optimization of the fulfillment industry will have robust effects on businesses and people. The increased amount and speed of deliveries has already resulted in a new retail store model coming into existence. This model is referred to as a showroom store or guide shop. The type of store is normally smaller, and more aesthetically pleasing than an average store. However, it only holds enough inventory to show the consumer every possible option needed to make a purchasing decision. The consumer then pays at the store, and the product is delivered to wherever they want. For people, technology will result in faster delivery of what you want, when you want it.

Amazon already offers 2-hour delivery in a lot of major cities. Soon you will have increased access to a greater selection of goods and services that will be delivered right to your door. In two to five years, consumers will even be able to seamlessly track their packages.

In theory, one would be able to go into a guide shop or go online, purchase an item, get a message when the warehouse receives and packs the order, watch a video of a robot that loaded it onto a delivery truck, and then track that truck to your door. The fulfillment industry will only get more and more efficient as companies start to embrace this technology and leverage the data to their advantage.

While there is a lot of positive development in the robotics and automation industry, there will also be a lot of disappointment in the coming years. There is a lot of hype in the robot world, but there is a fundamental gap in between what roboticists are building and what the people need. The amount of work that people are spending on purely academic exercises is wasted because a lot of it doesn't have a future. People need to think about the fundamental question behind robotics differently. Instead of approaching robotic development from a "build a robot that does..." standpoint, the industry needs to be saying, "What kind of robot can solve X problem?" While robots that can run, jump, and do a backflip are cool to see, a fundamental change in approach will help lead the industry to build more useful and needed robots in the future. ~

Bruce Welty is the Chairman at Locus Robotics Corporation. Locus Robotics provides an autonomous robot that streamlines warehouses and distribution centers to increase the efficiency of the fulfillment industry.

Source: Locus Robotics

Robots You Don't See

THE FUTURIST INSTITUTE

Robots You Don't See

Jason Schenker

- Chairman of The Futurist Institute -

When people think about robots, they are often thinking about anthropomorphic machines. They look like humans, they act like humans, and they move like humans. But that's not what a lot of robots look like that have significant value for our economy and society. Many of them look like boxes or have other simplified shapes, which optimize their functions to most efficiently counter every robot's biggest challenge: gravity.

Interestingly, some of the most important robotics and automation solutions operate behind the scenes. They are robots you don't see. They are in factories, warehouses, and distribution centers performing material handling functions within the supply chain, working hard to fulfill the promise of e-commerce. These unsung technological heroes of our increasingly digital and virtual economy are important now — and they will only become more critical and essential for the economy in the future.

Simply put: these robots and automated solutions are moving things around that would be cumbersome for humans to do and that will only become a bigger challenge over time.

In Q3 2017, e-commerce sales in the United States hit an all-time record high level of $115.3 billion. But that still represented just 9.1 percent of all U.S. retail sales. That percentage is going to rise significantly over time, and this will put increased burdens on the U.S. supply chain to fulfill what I like to call the promise of e-commerce.

E-Commerce Retail Sales in Q3 2017 at $115.3 Billion

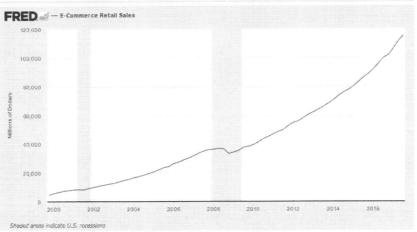

Shaded areas indicate U.S. recessions
Source: U.S. Bureau of the Census, FRED

FI THE FUTURIST INSTITUTE

I've often given presentations where I share the current percentage of retail sales that comes from e-commerce with an audience, and they almost always overestimate the percentage by a very wide margin. I've heard 30 percent, 40 percent, and higher.

Yet, e-commerce is still less than 10 percent of retail sales. It will get to those higher percentages in the future — but not without physical robots and automation solutions.

E-Commerce Retail Sales as a Percent of Total Sales in Q3 2017 at 9.1%

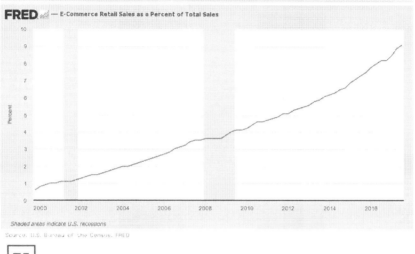

Shaded areas indicate U.S. recessions

Source: U.S. Bureau of the Census, FRED

FI THE FUTURIST INSTITUTE

Beyond The Supply Chain

There are other areas where robots you don't see are critically important, and 2018 is likely to be a big year for more rapid adoption of chatbots and other professional services automation tools that go beyond physical robotics and automation. The integration of software and hardware is also likely to accelerate. And software-defined automation, as Louis Borders notes in his essay, will be an important part of the process for moving robots forward.

Most importantly, 2018 needs to be a year in which we examine our lack of trust in robots, which is discussed in the essays from both Michael Walton and Daniel Theobold. And we need to recognize the fact that China is moving rapidly ahead in the field of robots and automation, which is something that Henrik Christensen focuses on in his essay.

If we don't address some of these core issues, we will hinder our own economic advancement, and the ramifications could be significantly negative. I've structured this chapter to start with these discussions of trust and China, and end with one of our most optimistic forward-looking pieces by Fred van Beuningen. But if we want to reap the benefits that robotics and automation offer, we need to move beyond our distrust of and aversion to physical robots and automated solutions.

As Michael Walton points out, "Robots and automation are not optional." The sooner we embrace this notion, the better off we will be — and the sooner we can usher in an improved state of conditions. ~

Jason Schenker
Chairman of The Futurist Institute
Editor of *The Robot and Automation Almanac*

Robots and Automation are Not Optional

Michael Walton

- Microsoft Industry Solutions Executive -

There's a big problem people have with robots. It's not fear, although the media and Hollywood might portray it that way. The big problem is more fundamental: it's a lack of trust. And the global perspectives are not equal when it comes to trusting robots. In 2018, this is likely to become more apparent, and this difference will begin to pose greater threats to our long-run economic potential.

The distrust of robots is largely native to North America and Europe, because countries like China are operating in different economic conditions, and they are subject to different values regarding human life and safety. In Europe and North America, there is a real concern that automation could obliterate jobs and entire fields. While this is likely overblown, the net impact of robots and automation for the job market remains a question. And there won't be much clarity on this point in 2018.

The year ahead is likely to be a year in which politically expedient anti-automation rhetoric taps into the unknown risks to the labor market and becomes a political lightning rod.

Meanwhile, our counterparts in Asia are experiencing unprecedented growth and physical demands that people cannot hope to sustain. While this is true in the U.S. supply chain, warehousing, and material handling industries, this is broadly true in Chinese manufacturing and a number of other areas of economic opportunity. While North Americans and Europeans worry about robots taking the jobs, the Chinese have no concern about robots taking jobs. In fact, the lack of workers in China is such a critical looming issue that the Chinese government has reversed its stance on the one-child policy. But until those yet-unborn children can become productive members of the Chinese economy, a solution is required. And that solution comes in the form of robots.

In North America and Europe, safety concerns, unions, and regulations are likely to remain a hindrance to more rapid adoption of robots. But in China, shipments and deployments of robots are likely to reach new all-time highs. While North Americans and Europeans debate the value of robots, the Chinese are moving full steam ahead.

And while robots are safer, people in North America and Europe refuse to believe. They refuse to trust. Of course, trust takes a long time to build, but currently people spend more time debating risks posed by robots, chatbots, and automation rather than monetizing the cost and time savings they offer.

An analogy I like to use when comparing the global dynamics of robot trust is dating and marriage. In Europe and North America, people view robots as dating. They see a trial-and-error process in which trust gradually builds and is continually tested. In this scenario, robots are viewed as optional. In China, robots are part of an arranged marriage for the economy. They are not optional. They are — like it or not — going to be part of the economic conditions going forward. People know they have to make it work, because this is set in stone. Hopefully, 2018 will be a year when more trust builds in North American manufacturing, supply chain, and retail parts of the economy, because we should be embracing a future with automation.

Our economic clock is ticking, and it's time to get serious — and marry those robots!

In all seriousness, people are more likely to buy a tube of toothpaste that a *person* covers with a piece of tape than to trust a robot that did it. And Western governments are the same way. North American and European governments won't accept robots doing 100 percent of anything. They, too, prefer the tape. This cumulative skepticism of the general populace, combined with government reactionaryism threatens to turn the United States from a country of tech pioneers into a wasteland of Luddites, while China takes the lead in robotics and automation.

A critical factor for the next few years will be a significant integration of cognitive services and artificial intelligence with robots. While this is emerging now — and is likely to be steadily increasing in 2018 — a broad adoption of these more integrated solutions for hardware and software is likely in the five- to ten-year window. Self-organizing bots, maintenance bots, and more comprehensively collaborative robotic and automation solutions are coming in the years ahead. This is a bit further out than 2018, but these worlds are coming together. And the West could lose out.

We need to learn to trust technology — before it's too late for our economy. ~

Michael Walton is a Director and Industry Solutions Executive at Microsoft where he is helping discrete manufacturing companies to digitally transform.

Robotics in China — The Next Frontier?

Henrik I. Christensen

- Contextual Robotics, UC San Diego -

Introduction

The robotics industry has seen significant growth since 2009. The consolidated annual growth rate (CAGR) has been 17% (*World Robotics*, 2017).[1] It is important to recognize that the growth rates are not uniform across the world. More than 50% of all industrial robots are sold in Asia. China is responsible for more than 2% of all robots sold. The Chinese market is also the fastest growing market. Over the past few years (2010-2016), the year-to-year growth in China has been close to 50%, which is impressive.

The reasons for the rapid growth in China are multi-faceted. China has a national strategy (Made in China 2025)[2] that outlines the ambition to become the manufacturing provider to the world by 2025.

Already today, 32% of all cars are manufactured in China. A challenge is to secure a predictable quality, which is the reason for adoption of robotics across the automotive sector. In comparison , the United States is already using robots extensively across the plate shop, welding, and paint shop for automotive manufacturing but less than 6% of all cars manufactured worldwide are made in the United States of America. The U.S. market is not growing, and it is considered mature. Electronics is another major sector with major growth. In particular, for high-end products, there is a significant penetration of robots on the manufacturing line. Most manufacturing of high-end cellphones is automated. People on the line will mainly perform inspection tasks to ensure homogenous and adequate quality of the final product.

Chinese Market Dynamics

More than half of the robots sold in China are made by foreign companies. In almost all cases, sales and service are performed by joint ventures that are based in China. Close to 15 years ago, ABB moved its robotics headquarter to China. During 2017 Midea acquired the German company KUKA to make it a Chinese company, but still with major production and R&D in Germany.

Today FANUC, KUKA, and ABB are responsible for 50% of the robot sales in China. In many cases, the installation of robots is performed by local systems-integration companies. Some of the integration companies slowly launch products of their own.

The company Siasun, which was sponsored by the Chinese Academy of Science, grew from systems integration to become the largest fully Chinese-owned provider of robot systems. Companies such as Siasin, Efort, GSK, and Estun are slowly emerging as Chinese robot providers. Siasun has already started to sell its products in the United States and Europe. Close to 20% of all robots sold in China today are manufactured by domestic companies.

An important indicator of the maturity of a particular vertical is the number of robots deployed per 10,000 workers. The rule of thumb for automotive is that one can deploy 1,000 robots per 10,000 workers, or a ratio of 1:10. We are seeing slightly higher numbers for Japan — about 1,400 — and on-target numbers for Germany, the United States, and South Korea. China, on the other hand, is only using about 300 robots per 10,000 workers, so they are still only at 1/3 of the utilization level expected for the car industry, despite today being the largest provider of cars in the world.

For general industries, the average is significantly lower. South Korea has about 450 robots per 10,000 workers, whereas Japan and Germany are around 300, and Sweden, the United States, and Taiwan are about 200. In comparison, China is around 50. The world average across all nations is 87 per 10,000. Consequently, China is far behind the average countries and very far behind modern manufacturing countries such as Germany, Japan, South Korea, and the United States.

The 50% annual growth numbers are thus not surprising. One would expect the growth to continue at least for a five- to ten-year period, which is also very much in line with the Made in China 2025 strategic plan.

Summary

The manufacturing market in China is booming. In parallel there is also tremendous growth in the service sector across drones (with companies such as DJI), logistics (Alibaba), and autonomous driving (such as Baidu and TuSimple). The broader service area will pick up significantly. We have already started to see simple products such as vacuum cleaners enter the US and European markets. There is no doubt we will see massive growth in Chinese robotics products across manufacturing and service applications over the next few years. It will be interesting to see how these efforts will grow, but also to see if US and Europe will try to counter these trends. ~

Notes

1. World Robotics 2017 - https://ifr.org/worldrobotics/, Brussels, Sept 2017.

2. Made in China 2025 - http://english.gov.cn/2016special/madeinchina2025/, Beijing, 2016.

Henrik Christensen is a roboticist and Professor of Computer Science in the Department of Computer Science and Engineering, at the UC San Diego Jacobs School of Engineering. He is also the director of the Institute for Contextual Robotics.

The Future is Built Upon the Past

Jayesh Mehta

- Marketing Manager of Transbotics -

Once upon a time, there was a factory, which long ago was successful, and the employees were paid well. As the factory grew more and more successful, more people were employed. Then one year, sales productivity went down, and with that so did the profitability. This went on for a few years, each year worse than the previous, but the factory kept hiring employees and paying them well, with the hope that fortune would change by throwing money and people at the problem. It did not. The employees were abundant and would do menial jobs to keep themselves busy. There was no motivation because they were doing tiresome, menial jobs. The factory was in danger of closing.

Then one day, the powers that be at the plant decided that they would need to automate their business, to make themselves more efficient, or to shut down the factory.

This would mean some employees would lose their jobs, but money would be saved for the ones that remained. They needed a solution and searched, and they found one: Automatic Guided Vehicles (AGVs). They purchased these in a last-ditch effort. The employees, as a group, were mad. They said, "You are taking our jobs away!" The politicians said, "You are hurting the economy! How will you replace the taxes your former employees used to pay?! We should tax your robots!" People were scared.

One year later, the factory was prosperous again. They had let go of 150 of their workforce but saved 850 others. By the second year of success, the company decided to build another factory, automate it the same way as the first, and hire the same size workforce as it currently had. So, although 150 people lost their jobs, the other factory hired those 150, plus an additional 700 people.

This little story is true. Technology does not have to be the scary monster in the closet. We don't need to fear change but embrace it. Technology has been going on for centuries. Knives were once made by scraping rocks, and then another, better process was discovered, and then another, and so on. The assembly line was a new process, and then the conveyor system improved upon this. Then pneumatic tools assisted the line worker. Forklifts were at one time a new technology, just as Automatic Guided Vehicles (some people will call them "autonomous," although not 100% accurate) were a new technology in the 1950s. Wait, what? Yes, the first AGV was installed in 1953. It has taken almost 65 years to get companies and people to start accepting them, and for some to try and tax them.

AGVs have come a long way from their inception. They used to only go wherever the metal tracks led them. Then they went by electric current, and then magnets, then laser target triangulation. And now by multiple means, such as cameras, laser bumpers, RFID tags, etc.

The question I am asked is why it took 25 years to go from tracks and wire to magnets, an additional 15 years to get to laser, and then another 20 or so to get to where we are now? There are a couple of reasons. The first I will explain with another example (please don't moan). When the modern computer was first built, it took up buildings to do simple problems. It took a long time to get to the personal desktops and then to the laptop, and now to the smartphones, each time getting smaller, less expensive, and more powerful. Each of these took a shorter time to be developed than its predecessor. Same thing with AGVs. But wait! What about the 20 years from laser to all the other types of navigation? That is the second reason. Safety, reliability, and repeatability. It didn't take the other navigation types 20 years from the laser targets to be introduced, just to become popular. To the outside world, a new technology seems to come out of nowhere, but to the manufacturers, it had to be in the works for years. In the material handling world, disruptive changes happen over time.

So, what is the next big thing for the coming year? Where will it come from? How will it affect the industry? I believe in the next year that the big thing will be AGVs in general, but more specifically the types of sensors and navigation for the AGVs. As I said earlier, there really is nothing "new" within the industry, just what is released and improved upon.

I'll break down individually what I think will be hot tech for the next year (and possibly beyond).

I spoke briefly on the navigation earlier, and you probably were thinking, "Okay, where is he going with this?" I wanted to get the history out of the way so we can start here with the new. There were a lot of new navigation methods released in the last few years, but most are not that reliable. That doesn't mean that they are not worth pursuing, just that they need time to develop. The most promising navigation medium is "Feature" or "Natural" ...I'll just call it Feature because I don't want to write both names down constantly. This was released about two to three years ago, but with limitations. It had to "see" something in over 50% of its view. This was great for smaller rooms (typically under 50 meters square), but it also needed to have over 50% unchanged, making this a bad application for warehouses, deep lane storage, and larger production areas. Some also had a large margin of error, of up to ±6", which can be problematic.

Despite these limitations, some companies are willing to try the technology, which helps fund research, so improvements can be made. Also, with its popularity rising, other companies that don't offer this technology and component manufacturers are trying to develop this and make it better than what is currently out on the market. It is now a race, with a lot of time and money going into it to see who can put out the best product and grab the first large share of the market.

The automotive industry has started this "autonomous" car thing with good intentions, and it will be a good thing once it gets working on a regular basis in every condition. But for now, the industry is dumping millions into research, and not all within the companies themselves, but also to outside vendors. These outside vendors have been creating safety sensors for high speeds and to help identify what the object in front of it is. AGVs don't run that fast, for safety and to prevent product damage. That doesn't mean that they can't utilize some of the sensing technology that the automotive industry has helped develop. The manufacturing and warehouse industries will start seeing AGVs that can tell the difference between a pillar, a person, a manual vehicle, a pallet, and a piece of scrap paper. This will help AGVs become more efficient than they are currently. For example, if a piece of paper is on the ground, currently the AGV will stop until it is moved out of its laser bumper sensor. With the new tech, it will know it is just a piece of paper and continue moving, reducing stop times. In the same vein, the vehicle would also know how fast an object is moving, so if it happens to be following a person, it will know not to speed up and slow down, causing wear on the motor, wheels, and brakes. It would match the speed.

I believe this coming year will see great strides in these technologies and help set up the AGV as a hot commodity for the next few years, until the next big thing comes...we're watching you, drones! ~

Jayesh Mehta is Transbotics' Marketing Manager and co-author of *A Rogue's Guide to Acquisition: Principles from the Final Frontier.*

The Impact of Automation on the Supply Chain

Daniel Stanton

- President of SecureMarking -

2018 promises to be an important year for the adoption of automation in the supply chain. From robotic warehouse workers to autonomous vehicles, to automated order fulfillment systems, new technologies are starting to transform the way that we engineer and manage our end-to-end processes. But while automation is enabling the emergence of new legitimate business models, it is also creating new opportunities for counterfeiting, theft, and other illicit businesses.

One useful way to monitor the rate of technology adoption in the supply chain is by studying the evolution of common processes using the Information Value Chain.

Most supply chain process frameworks, including the Supply Chain Operations Reference (SCOR) Model, define supply chains according to six top-level processes:

Plan

Source

Make

Deliver

Return

Enable

Each of these processes can be broken down into a number of subprocesses, and each of the subprocesses includes tasks that involve some combination of human effort and automation.

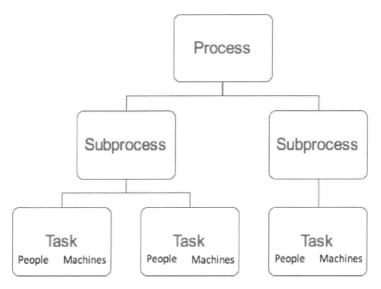

Graphic: Provided by Permission from Daniel Stanton

The effective division of labor between people and machines can be illustrated with the Information Value Chain, shown below. Machines handle the heavy lifting (in terms of both physical products and large amounts of data), and people focus on activities that involve intuition, judgement, and creativity.

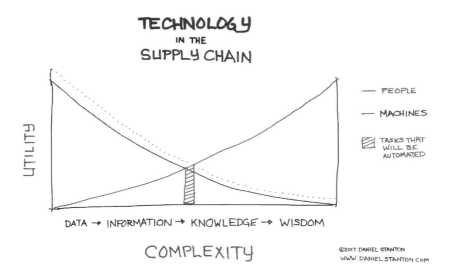

Graphic: Provided by Permission from Daniel Stanton

As in prior years, each of the top-level processes has seen a clear upward shift in the technology curve during 2017. In other words, technology has continued to advance and take over tasks that were previously being done by people:

Plan: Planning supply chain networks is a computationally complex task that has long been the domain of sophisticated software. It includes selecting sites for facilities, selecting transportation modes, and developing plans to balance supply with demand. In 2017, the most important single technology emerging in this space was blockchain, which promises to reduce the guesswork of planning by allowing the partners in a supply chain to share more information, more securely. The reliability and transparency of blockchain effectively automate the human task of "trust" between supply chain partners.

Source: Procure-to-Pay processes have largely been automated as a part of the wave to implement Enterprise Resource Management systems. The jobs that continue to require human intervention are activities like supplier relationship management. Even in these areas, there continue to be new tools emerging that streamline and automate key processes, such as ThirdPartyTrust, which monitors supplier certifications and credentials.

Make: The market for industrial robots has continued to grow. Part of this is a continuing trend of big, heavy, expensive machines that can outwork and outpace humans. But in 2017 we've seen smaller, faster, more collaborative robots become more common. For example, Amazon Robotics (formerly known as Kiva) is now common on TV ads, and new entrants like Locus Robotics are bringing in simple, inexpensive products at the bottom of the market. Walmart has begun to deploy drones with RFID readers to perform inventory counts in its distribution centers.

Also important in this process is the growth of additive manufacturing, allowing companies to do rapid prototyping and distributed manufacturing.

Deliver: In the third quarter of 2016, Otto demonstrated an autonomous truck by hauling a shipment of beer 120 miles across Colorado, and Amazon Prime Air made its first delivery in Cambridge, England. Late in 2017, May Mobility launched its autonomous taxis in Detroit. Several companies are expected to demonstrate autonomous cargo ships in the next year. From long-haul intercontinental freight to last-mile parcel delivery, 2018 will be filled with pilot projects and new market entrants for cargo in every mode of transportation.

Return: Returns is one area that tends to require a lot of judgment and therefore human intervention. Inspecting products to determine their condition and making a decision about disposition are still labor-intensive tasks. Returns have become a major cost element in e-commerce supply chains, often involving as much as 15% of total sales. Returns are also a potential vulnerability, as customers may return stolen or counterfeit products. Technologies that provide manufacturers and retailers with the ability to authenticate products and track their provenance are well established in regulated markets such as pharmaceuticals, tobacco, and alcohol products. But they are beginning to see adoption in new markets such as fashion brands, electronics, and military hardware.

Enable: All of the technologies discussed for the other SCOR processes are enabling processes. The automation of the Enable process basically means having one machine select another machine to perform a given supply chain process. Clear-cut examples of this are not easy to find today but are probably not far out on the horizon. For example, we can imagine a point in the future where a customer would issue a purchase order to Acme Inc. for 100 widgets, and Acme's computer system would automatically allocate and assign the production order to 10 different additive manufacturing facilities based on their capabilities and available capacity.

The adoption of these technologies is leading to some major trends that I expect will become more apparent in 2018:

Convergence of Planning, Execution, and Visibility
In the past, there has often been a disconnection between planning, execution, and visibility in supply chain processes. For example, one system would be used for inventory planning, another for managing order fulfillment, and a third for reporting on current inventory levels. Breaking these functions apart made the jobs more manageable for human analysts to perform. But with the rise of automation, these functions are converging. This is not yet a well-defined software category, but Elementum appears to be the pioneer that others are likely to copy in 2018.

Emergence of "Supply Chain" as a Business Model

In many cases, the adoption of supply chain technologies is being driven by cost-saving opportunities. But in other cases, it is a response to an inability to recruit human workers, or an inability of humans to perform to levels required to meet customers' expectations. The single biggest driver of these rising customer expectations has been Amazon. And what gives Amazon such dominance is the combination of a strong retail presence combined with information technology and logistics infrastructure. In other words, Amazon is a supply chain company. It appears that Amazon may be the new paradigm for retail supply chains, in which case the years ahead are likely to see consolidation between otherwise unrelated companies. For example, while Walmart, UPS, and Microsoft are all competing against Amazon on a variety of fronts, through a merger they might form a new supply chain company that could compete more effectively.

Growing Threat of Counterfeiting

In 2017 there were a number of U.S. recalls of food, medicine, automobiles and other unsafe products. At the same time, a U.N. study reported that 2% of all goods traded internationally are counterfeit. Increasingly, supply chain managers are becoming responsible for authenticating products throughout their lifecycle, and for monitoring how, when, where, and under what conditions they are moved and stored. The only effective way to manage track and trace and ensure the authenticity of goods is by securely marking and tracking them at the individual item level. ~

Daniel Stanton is the President of SecureMarking. As a Supply Chain Expert, Daniel is working to provide a solution to the growing threat of counterfeit, black market, and grey market products in global supply chains.

Robotics in the Warehouse: The Autonomous Transportation Revolution Comes Indoors

Lou Micheletto

- Manager of Integrated Solutions, Yale Materials Handling Corporation -

Gone are the days when self-driving cars were confined to science fiction and bold predictions of the future. With major automakers and technology giants alike pursuing self-driving cars, 30 companies had permits in California to test autonomous vehicles on the road as of April 2017.

The supply chain, specifically inside critical manufacturing and distribution facilities, is another proving ground for the autonomous transportation revolution. While these environments have used a previous generation of technology for decades, advances in mobile robotics can open a new frontier of greater capability and performance.

Automated guided vehicles (AGVs) first gained widespread adoption in the 1970s to automate basic point-to-point load transportation, relying on extra infrastructure, such as ground wires, tape, magnets. or reflectors to navigate rigid, prescribed paths. In the event of an obstruction, AGVs would stop and beep until removal of the obstruction, and any adjustment to facility layout required reinstalling guidance infrastructure.

Today, advances in navigation technology and software can enable standard lift trucks to function without an operator, producing a robotic lift truck solution that sheds the limitations of traditional AGVs and offers far greater flexibility. Many of these solutions use LiDAR — a laser-based navigation technology that produces a two-dimensional view of the facility, measuring the building structure by looking for hard features like columns, walls, and racking. In effect, it builds a map of the facility, collecting more than 50,000 data points every two seconds to self-locate and navigate in real time. This mapping offers great strides forward in terms of flexibility, ease of use, and productivity, allowing robotic lift trucks to easily respond to layout changes and re-route in the event of unexpected obstructions.

But as advanced mobile robotic lift trucks establish themselves as viable solutions, what demands in manufacturing and distribution drive adoption in 2018 and beyond?

Labor Challenges

According to the Bureau of Labor Statistics, the turnover rate for warehouse workers is 36 percent. And other industry sources peg the cost of filling vacant positions at anywhere from 25 to 150 percent of an employee's annual salary. Midway through 2017, U.S. warehouses had 600,000 unfilled jobs, and Goldman Sachs economist Daan Struyven reported findings that suggest the labor market has already slightly overshot full employment. Combine all that with a rising minimum wage, and labor is becoming increasingly expensive, harder to find, and more difficult to hold onto.

Operations must re-evaluate how to deploy labor and implement technology. Using automated solutions for basic point-to-point materials transport tasks frees employees to advance to more engaging, meaningful positions — helping address employee turnover and labor-sourcing challenges.

Adopting robotic lift trucks can also help simplify tasks reserved for employees. Picking e-commerce orders is a very labor-intensive process, with employees walking a multitude of storage aisles to retrieve merchandise and pack orders. However, implementing a goods-to-operator fulfillment workflow reverses this travel by bringing products to employees. Pickers do not need to know specific inventory storage locations, nor must they spend time walking between aisles. Instead, they can focus on picking and packing orders as efficiently as possible. Similarly, operations could reassign high-performing lift truck operators to manage multiple robotic units from a tablet — elevating their roles and eliminating less-valuable time spent in transit.

The Uptime Imperative

A study by Information Technology Intelligence Consulting found that 98 percent of organizations surveyed across several industries say a single hour of downtime costs over $100,000. Even minor interruptions or unplanned outages can cause immediate harm to the bottom line, like missed shipments and lost production time. Trickle-down effects can include overtime and rush expenses as operations attempt to catch up, and lost customers due to delayed shipments.

Automation offers a reliable solution, capable of minimizing the risk of downtime and unexpected delays. A robotic lift truck doesn't need sick days or breaks and only stops for battery charging, replacement, and maintenance. They always follow the rules of the road, which can reduce impacts and unplanned interruptions compared to manned equipment while improving equipment longevity. Most importantly, standard lift trucks outfitted with robotic technology can be serviced by the same local personnel who already handle lift truck fleet maintenance, avoiding the downtime and expenses of specialized technical resources.

Tight Schedules Drive Continuous Optimization

Supply chains are squeezed to do more with less as they strive to meet increasingly strict delivery commitments at minimal cost. Services that were once considered perks, like two-day and next-day delivery, are now an expectation. As such, supply chains engage in a cycle of continuous process optimization and allow for smaller margins of error than ever before.

These operations require technology that can help prevent minor issues from affecting an entire fleet and derailing throughput. For example, the software that manages robotic lift trucks finds the most efficient routes and responds in real time to obstructions and shifting traffic volumes, directing units to alternative paths as needed. If one robotic lift truck encounters a delay-causing obstruction, it can inform other units, so they can find alternate routes. This is a huge leap from traditional AGVs, which are not only restricted by route infrastructure but also lack onboard intelligence to find alternative paths, much less inform other units.

To 2018 and Beyond

Underlying trends in labor, consumer preferences, and technology are pushing manufacturing and distribution facilities to adopt mobile robotics. But these immediate operational realities are not the only factors at play. Industry 4.0 and the Industrial Internet of Things are set to usher in a new era of connected devices and machine communications. As businesses look to capitalize on these developments, mobile robotics gathering data and communicating with a central repository can play a valuable role. ~

Lou Micheletto is the Manager of Integrated Solutions at Yale Materials Handling Corporation. Lou has more than 40 years of experience in the materials handling industry, with a background in product design and sales. He leads the company's automation initiatives.

Source: Yale Materials Handling Corporation

Supply Chain Automation: Humans Will Always be *On* the Loop

Robert Handfield

- Professor at N.C. State University and Author -

When it comes to automation in the supply chain, there is an awful lot of speculation, an awful lot of hype, and a whole lot of misunderstandings. It is not uncommon to hear of writers discussing the role of machines versus humans and their relationship, especially in the context of supply chain activities, including warehouse operations, embedded sensors, procurement, transportation, and financial supply chains.

Scenarios are being painted of computers operating in a real-time environment, using blockchain to process transactions, relying on the Internet of Things to order goods and services, which are delivered by drones to consumers' homes. The reality of this scenario is far-fetched indeed, primarily because futurists are dramatically underestimating the degree to which people and organizations must change to deal with these technologies.

A similar level of euphoria existed 17 years ago, when people started playing against computers in chess. For instance, recall the level of discomfort many readers may have experienced when computers were playing against and beating Grand Chess Masters in the 1990-2000 period, described by Garry Kasparov in his book *Deep Thinking*:

> The growth of machines from chess beginners to Grandmasters is also a progression that is being repeated by countless AI projects around the world. AI products tend to evolve from laughably weak to interesting but feeble, then to artificial but useful, and finally to transcendent and superior to humans...Overestimating the potential upside of every new sign of tech progress is as common as downplaying the downsides. It's easy to let our imaginations run wild with how any new development is going to change everything practically overnight. Human nature is simply out of sync with the nature of technological development. We see progress as linear, a straight line of improvement. In reality, this is only true with mature technologies that have been developed and deployed. We expect linear progress, but what we get are years of setbacks and maturation. Then the right technologies combine, or a critical mass is reached and boom, it takes off vertically for a while until it reaches the mature phase and levels off.

Kasparov also notes that it is important to recognize the role that machines have in AI and the importance of humans "on the loop" through the use of "Moravec's paradox."

In 1988 the roboticist Hans Moravec wrote, "It is comparatively easy to make computers exhibit adult level performance on intelligence tests or playing checkers, and difficult or impossible to give them the skills of a one-year-old when it comes to perception and mobility. Computers are very good at chess calculation, which is the part humans have the most trouble with. Computers are poor at recognizing patterns and making analogical evaluations, a human strength."

Or as Bill Gates stated in his axiom: "We always overestimate the change that will occur in the next two years and underestimate the change that will occur in the next ten."

The same argument exists for the role of automated vehicles and drone deliveries in transportation logistics. I recently had the opportunity to listen to Missy Cummings, a professor in the Pratt School of Engineering at Duke University. Missy shared some insights on automation in the logistics sector, at a conference that was awash with discussions on blockchain, digitization, and new technology in logistics. Missy noted that, "I was one of the US Navy's first female fighter pilots, and Yossi Sheffi from MIT, who I worked with (who was also a fighter pilot with the Israeli air force). One of the toughest things we did was land a jet fighter on an aircraft carrier — and it opened my eyes to the future. One of the things you learn quickly in the landing is to turn the computer on and keep your hands off the controls. The computer handles it, but you may take over only if it gets tricky. In fact, when you take off, the pilot has to show their hands in the air and show them to everybody, to reassure everyone that you are *not* touching anything.

And only then do they fire you off the carrier. If you touch anything, a human can create pilot-induced oscillations off the front of an aircraft carrier, and there were many fatalities in this case. This started me thinking: 'I am so good I am not allowed to touch anything!' But where is logistics going, if pilots can't touch the planes on takeoff?"

Professor Cummings then gave the audience a brief history of humans and automation. At the turn of the century, horse and buggies were entrenched as the primary form of road transportation. There was a real battle among people, many who insisted we would never get rid of the horse and buggy. Not only have we moved to cars, but we are now moving to driverless cars. Next, on Wall Street — the human interactions on the trading floor have been replaced by computers. Most of the financial industry today (FinTech) is done by computers. In aviation — the old jets are replaced by UAV drones with no pilots. So, we have achieved that level of automation technologically, but we won't have full replacement of pilots, primarily due to regulatory issues.

Cummings works on human supervisory control. A human operator is trying to control a computer vehicle, and it mediates the commands that are interpreted by actuators into a drone. So humans sit on the loop (not *in* the loop) — as they are not directly controlling the systems. We have sensors and displays that are used to balance humans in the automation. But should humans be given stop buttons, or what is the nature of their interactions?

Computers work best in automation when involved in complex, time-pressured high-risk domains, requiring embedded automation/autonomy with human supervision. Think of mining trucks in Australia or nuclear power grids — there is a lot going on and a lot of information being taken in. Humans are increasingly moving from the mechanics of operating equipment to the cognitive interpretation of what is happening. Every commercial aircraft is basically an automated drone today. The pilot could be doing their job from a Starbucks coffee shop! Pilots touch the stick about 3.5 minutes out of an entire flight, and that is on takeoff only. That is due to the regulatory structure — and it will soon go away. Even NASA lets the astronauts have too much control, and it resulted in suboptimal landings for the first few trips to the moon. Today, humans are only involved in case something goes wrong. If I can get rid of the human capital, labor and retirement costs will go down. But there must be a balancing act somewhere. In China, we see cheap labor, and even then, they have decided to start investing in robots. There is a balance between humans and machines, and if I can automate all these jobs, is there a potential human cost?

So how does automation work? Automation has to be able to assess the situation and understand what is going to happen — and it has to focus on skills-based reasoning first. Over time, as it learns more and more situations, that leads to expertise. But humans are the only machines that truly have the highest level of expertise.

Of the transportation domains, rail has the lowest uncertainty, as you are traveling in a one-dimensional vehicle, which is akin to skill-based reasoning. Next up in terms of uncertainty are drones, which are the easiest to automate. Drone technologies are mature, have been around for 30+ years, and are rules based. Airplanes are also rules based, and because there are three degrees of freedom, there is also more room for error and less likelihood of head-on collisions.

But the most dangerous — and the most uncertain — technologies are driverless vehicles. There are many, many more uncertain conditions occurring, and the trajectory is only in two dimensions, so there is much less room for error, compared to aircraft, which have three dimensions. Driverless cars require expert knowledge and reasoning. "Are there kids playing soccer; what about animals, and how do I react?" That is pure expert-based reasoning, and we are simply not there yet with computers to handle this level of uncertainty.

Unfortunately, these technologies are being introduced in reverse!

We are primitive in automated rail. There are third-world countries that have more successful and more advanced automated rail systems. Drones are coming very slowly, with FedEx and UPS moving slowly, but there are no technological reasons why they will not evolve. There are some sociotechnical concerns, however. But driverless cars!

Dr. Cummings is working with groups in Washington, D.C. on proposed legislation to put one million driverless cars on the roads in the next year. In her opinion, "We are not ready for it, but that is the power of lobbying markets. And car companies are cutting deals with Uber, Lyft — and there is a big business out there, but it is much too early in the evolution of technology for driverless vehicles."

Based on my research and ongoing discussions I've had with executives, here are the three big ways paradigm shifts will impact events going into 2018:

The first is the importance of effectively defining organizational data governance. Analytics are the product of data — so if you don't trust your data, what does that say about your analytics? Ultimately, robots are going to have to rely on data produced by the Internet of Things and other sources, so getting your data in order is essential. Organizations that get their data governance in order will have first-mover advantage.

Next, there is a need to really understand how data will be used in your organization and among parties in the supply chain. This will take some time to work through all the channels. Sharing data sounds great, except to a lawyer! The intent is to engage people to share "what decisions could you make if you had this specific information?" as a means to identify how to design not only the user interfaces but the timeliness and specific types of data required for these decisions.

To enable machine-based learning and automation, machines will have to learn from people how they use data to make decisions. Being able to see what "experts" are doing in terms of "best practices" can help establish the essential features of the data platforms and interface mechanisms. You have to train machines for them to operate — and new technologies will roll out that will follow humans like puppy dogs and learn from them!

The third challenge is how to roll out the digital automated economy among supply chain partners, especially getting people to share information, and to be honest and forthright in how they share data. How do you incentivize partners in the supply chain to be honest and share their issues and share their data? How can you trust them to trigger orders, without having to ask them and check on them? How do we reward people who go the extra mile and use smart contracts through emerging technologies such as blockchain? These and many other questions are more interesting to consider than how fast the technology will be adopted. Organizations that find ways to drive collaborative data sharing and get around the problems will also get the early-mover advantage.

The cultural transformation that will accompany these changes will be significant. Using the example of driving and "trusting" the Waze app to take one down the right street to avoid traffic, participants noted that Deere managers will have a hard time trusting new systems and following their guidance. The change -management portion of analytics will be a bigger inhibitor than the software and hardware changes.

In my opinion, machine-based learning and cognitive analytics will likely be adapted in a slow but steady rate in 2018, with the real shifts starting to occur in 2019 and 2020.

In the meantime, let's think about automating more trains! ~

Robert Handfield is the Executive Director of the North Carolina State University's Supply Chain Resource Cooperative. The SCRC is a university-industry partnership dedicated to advancing the supply chain industry and the professionalism of its practitioners.

Software-Defined Automation

Louis Borders

- Founder of HDS Global -

The industrial automation sector is booming. And because manufacturers and integrators are enjoying record years, one might overlook the disruptive transformation underway.

Software, to paraphrase Marc Andreessen, is beginning to eat the mechanical automation industry; new players are primed to dominate, and many current players will stagnate or cease to exist.

Despite the boom, the industry is ripe for disruption because large industrial customers are unhappy with today's systems. Custom-designed systems have two- and three-year lead times. Automation is not reducing labor fast enough to offset current labor shortages, especially in distribution. Mainframe-like monolithic solutions have poor ROI, and their rigidity can't keep pace with accelerating demand for versatility.

Software Will Dominate the Automation Industry

Today's hardware-first industry is transitioning to software-first. The mobile phone industry is a useful mental model that foreshadows the automation industry transformation. While the hardware in today's phones is remarkable, the software apps (Android, iOS, Facebook) dominate. Today's industry leaders are software companies, and none were in the phone industry before the transition to smartphones.

Bye-Bye, Conveyors[1]

Conveyor system: Monolithic with poor utilization — some lanes jammed; most empty.

Swarm of mobile robots: Diminutive with 100% utilization.

The elimination of conveyors is a key aspect of the industrial automation transition. Today, conveyors are the mainstay of large-scale automation systems. Tomorrow, the system mainstay will be mobile robots.

Mobile robots are to conveyors what PCs are to mainframes. Mobile robots are slow and expensive today, but they will be fast, cheap, and smart tomorrow. Their sensor-rich, 360-degree vision makes them safe in human environments.

Swarm-based software lets each mobile robot operate as an independent agent adhering to a simple rule set, which results in an astounding emergent collective behavior. Scaling is as simple as ordering more mobile robots. Installation time is literally 24 hours versus a year or more for a large conveyor-based installation. New workflows can be programmed with no hardware changes. Just as the mainstay of today's largest data centers are PCs, so the mainstay of tomorrow's largest industrial automation systems will be mobile robots.

Hundreds of test systems were launched in 2017. In 2018 there will be a few large-scale installations and a small but ominous slowdown in orders for large, conveyor-based, monolithic ASRS (automated storage and retrieval) installations.

Making the Transition

Individuals and companies can take specific actions to use this transition as an opportunity for advancement. Actions can be categorized along two paths — technical design and strategic/career planning — and several of these are highlighted below.

The intent is to trigger your own ideas for the many ways to use this transition as a once-in-a-generation opportunity.

Winning Through Technical Design

Focus design toward autonomous vehicles and articulated robots:

- A **software-first** mantra leads to simpler, general-purpose hardware design. If, in a design process, you have the thought that this simplified hardware design means the software is much more difficult, you are on a good path.

- For large systems purchased by industrial customers, strive for an **integrated hardware/software** solution. Model your design strategy after the Apple juggernaut of combining hardware and software and designing both at the same time.

- Recognize the power and price points of **consumer electronics** as highly viable industrial components. This carries over to software. For example, an inexpensive, powerful gaming engine platform is, for some use cases, better than industrial simulation apps. It's the total investment in consumer electronics that causes this surprising outcome. Billions are being invested in cameras and game engine platforms, but the cost per user is pennies.

- Use the software technology that operates **printers as a model for operating complex equipment**. A driver is written to sit between the complex system and the printer. The driver consists of simple building blocks or instructions sets. The complex system can get the printer to do anything and advance the output of the printer for years without changing the driver code.

In 2018, the first few big winners among the new robotics companies will emerge.

Winning Through Strategic/Career Design

- Seek companies and business opportunities that are not dependent on technologies that will be in decline. A key indicator is the **size and quality of the software effort** in a company. Is the company moving to a SaaS model? What is the size and caliber of its software team?

- **Study AI.** It's the real deal. Forty years ago, there was a great buzz about relational databases. Today, relational database technology is deeply embedded and pervasive in all major computers systems. Today, AI is on that same path. It is so powerful. It will make software smart and less intolerant — and it will reduce the great coding burden on all companies.

- Move to **product-focused** companies, not those with customized systems. The modern approach is to productize and configure rather than customize. Industrial customers are not happy with the heavy burden of managed software, while at the same time, paying hundreds of millions of dollars each for these large software applications, such as an ERP. By 2018, many large companies will likely be working to pull the plug on their ERPs.

I hope I've provided some insight into the future of the industrial automation world. If the transformation occurs as predicted, the next few years in automation will be fun and exciting for those leading the transformation. ~

Note

1. Image Source (Left): AdobeStock GraphicCompressor. Image Source (Right): Mercury Startups.

Louis Borders is a Founder of Borders Books, Mercury Startups, and HDS Global.

Cozying Up to the Last Mile

David Schwebel

- Senior Director of Swisslog Logistics Inc. -

Think back to 1910-1920 Sears Roebuck: the catalog, the department store, how you got your orders. It was a telephone storefront; everyone had one of those catalogs at home, and if you were in an urban environment, you could pick and choose from the catalog in the morning and pick up your order in a couple of days at the local Sears Department Store.

In the 1970s and 1980s, our industry realized it could achieve economies of scale by moving the warehouses closer to the manufacturing centers and inbound ports. Shop in store; have it delivered in a week at home. Today you can sit at your favorite coffee spot, order on your phone, and it is sent to your home the same or next day. Tomorrow, we are transitioning back to the urban environment — to get closer to the consumer, to shrink the distance between decision and consumption. That's what the last mile is all about: it is getting as close as possible to the end user.

Now it has already begun to turn into, while sitting at the coffee shop, my wife reminds me we need to pick up diapers on the way home. Instead of changing my afternoon plans, I have the near-instantaneous ability to accomplish my newly acquired goal of keeping my child dry at night — open my Amazon app, type "dia" (don't worry, the autocomplete algorithm kicks in, saving me even more thought time), completes to "diapers," picks up on my last order of diapers (remembering that I did this same task a month ago), and with one-click ordering, my fulfillment package is on its way. Funny thing is, the consumer doesn't care where it comes from — from a regional center three states away, an urban fulfillment center 30 miles away, or even the corner store only three miles away — just as long as my assumption of arrival on time and in full is achieved.

Fulfillment has come full circle. Last mile is really about distance, speed, and consumer perceptions — all of which the consumer is willing to pay an increased price for. Real, frictionless convenience.

Take an example: Deliv.co (http://www.deliv.co/)
It goes into an urban environment and brings together three major components of the last mile — all within the same ecosystem:
1. Retailers that want to do last-mile delivery.
2. A fulfillment center, which could be a reclaimed office location, old post office, or even a former furniture store.
3. A local workforce that knows the ins and outs of the unique city environment.

Deliv.co is a tech company. That's all it wants to be, yet the genius is in that it wants to get these players connected and use its multi-city contacts to the benefit of all companies in its ecosystem.

And just like that, a new(ish) fulfillment channel is born, providing regional and nationwide merchants the ability to access the last-mile fulfillment arena without having to create the capital-intensive infrastructure themselves.

Simply put: If each company in an urban environment that could afford to do a last-mile fulfillment method did so on its own, it would rapidly strain an already congested travel network — eroding consumer confidence in the fulfillment channel. So, Deliv.co is combining the ebbs and flows of capacity needs from all of these regional and hyperlocal businesses into a single fulfillment model — more efficient and effective than having 100s of these last-mile companies all competing against each other.

Top-named retailers are already on here:
http://www.deliv.co/participating-retailers/

What do the retailers want? They're trying to provide same-day packages. They want to bring their unique services and brand moments directly to the consumer in a physical way. Concierge processes and retail stores turned into showrooms, where stylists give customers the feeling of being pampered onsite and at home.

They take your measurements physically and now digitally, the backroom tailors already begin cutting before you even leave the store or set your phone back down, and now on your way home, the delivery people white-glove deliver it to your home at a time you select. It's all about bringing that moment of experience to the customer. A moment of wow — it ties your brand together with a desired positive experience. It encompasses all of it. And the consumer wants it a second time. And a third time.

And they will blog, Instagram, Facebook, and tweet about it.
You can't pay enough for that positive press.

Let's look at another example: did you know that the last-mile, each-fulfillment, hand-delivered process has been done in urban environments for the past 30-40-50 years? Restaurant daily replenishment. The fresh produce shows up in a van; the chef walks out of the restaurant and chooses the freshest and best vegetables that you'll be dining on that evening. You're bringing the goods to the individual for them to make the final decisions. Now it's a curated decision.

OK, one more example: Let's look at Blue Apron, an internet recipe-delivery service, by monthly subscription. We've trained the consumer public that smaller portions are the best way to go, because there's less waste in the system. Don't buy that 20 pounds of apples; just enjoy the best two apples now — it's all you need for tonight's recipe anyway. We desire the deeply-discounted prices, yet we really only want smaller quantities when we need them. The leftovers just get in the way.

It's all about location, location, and speed.

We're translating that digital desire into how we shop. Back in the 1990s, the grocer would bring the items to me, and if I were physically available, I could get the pick of the crop. Today? I'm in the coffee shop, in another state, and I realize I forgot my deodorant. Now, I want this individual item (the each) and I need it, and I'm willing to potentially pay an upcharge for the item to have it delivered to me, today, as close to me as possible, wherever I happen to be later that afternoon.

Granted, there is some price elasticity that's going on. If I usually pay $2 for that deodorant, and you want to charge me $10 to have it delivered, same day, to my office, I'm probably going to say no. Yet, if you say $2 with a $1.50 delivery fee — *Sold*! I can't push that buy button fast enough. That's where last-mile/urban fulfillment centers come into play — providing a fulfillment channel that *can* take advantage of economies of scale to provide a local market service the consumer is willing to pay for.

And we can't forget that we live in a global economy.

Last mile has now gotten to the point where we realize that sustainability matters tremendously. The great power that comes from this is, instead of me going to a Costco and buying a 70-pound bag of rice or a four-gallon jug of ketchup, I can turn around and buy smaller quantities and have them delivered to me, not in a cardboard box but in a shopping bag. Or to a locker outside the house.

I'm now forgoing the millions upon millions of tons of wasted cardboard I'm not buying in 50-pound bulk bags; I'm buying what I need and consuming when I need it. There are traditional existing companies, and then there are new entrepreneurial "pure e-commerce" companies paving the way.

Traditional companies are having to retrofit last-mile into their facilities. They already have the traditional warehouses, regional distribution centers, forward-deployment/fulfillment centers — all are about the pallet and the case.

Now the rise of the "fulfilled each": urban distribution centers could be in the back half of a local retail store, and the front half is showroom or pickup white-glove service. How do you add urban fulfillment and last mile to existing companies?

New entrepreneurial pure e-commerce companies are intriguing: they usually have their manufacturing centers in Asia or other offshore locations. They bring things in via ocean transport and add a break-bulk facility or import facility where they separate the goods into pallet quantities that they send to Amazon, which does the picking.

There is also an emerging concept of bringing your user into the last mile, because that's store foot traffic, and that's good for upselling. Track-and-trace or secure transport of goods to the consumer. The concern has always been that for some items, like laptops or liquor, you can't just leave them on the doorstep.

There are ways of continuing the track-and-trace or chain-of-evidence trail. Either draw customers into the store, or have them order it online and become part of the last mile by going through the drive-through for the local fulfillment center to pick up their package on the way home.

The quintessential last mile is about bringing the customer into the logistics. We've been teaching people to be part of the last mile, to be part of brands, to be sustainable, to be part of the value chain for the past 20 years.

They receive a text that their pizza is out of the oven and ready for collection.

Now, we now need to teach them to be part of the physical supply chain as well. Now it's all about speed and getting it when I need it and where I need it. ~

David Schwebel is a Senior Director at Swisslog Warehouse & Distribution Solutions, where he helps to develop and deliver automation solutions for forward-thinking health systems, warehouses, and distribution centers.

Acceptance and Industry 4.0 Drive Enterprise Deployments

Laura McConney

- Marketing Specialist at JBT -

Over the past ten years, technology has transformed the entire world. In more developed countries, it's hard to find people not looking at their devices. Meanwhile, elsewhere, mobile money, like Kenya's M-Pesa, has remodeled economies and the way that we think about them.

It should come as no surprise, then, that technology is dramatically altering manufacturing as well. In fact, some completely autonomous facilities already exist, and the demand for technology that makes such facilities work is higher than ever. Automated equipment, like robotic arms and automated guided vehicles (AGVs), increase efficiency, generate savings, and improve safety throughout the industry. Because of these benefits, we have seen companies adopt automation, in particular AGVs, at a remarkable rate over the past several years.

We expect this trend to continue and for organizations to invest in AGVs on an even broader scale. For example, at JBT, we offer customers an enterprise-wide deployment program, which enables clients to install our vehicles in their facilities simultaneously around the world. Companies have already begun forming these types of partnerships with AGV providers, and we are anticipating that enterprise-wide AGV investment will gain significant traction in 2018 and beyond.

While AGVs have been around since the mid-20th century, companies and individuals feared them for a long time. Without exposure to automated equipment like AGVs, many found them untrustworthy and unsafe. Automation was not something that they encountered in everyday life and, as a result, was greatly misunderstood. However, because exposure to technology in general has grown, in particular with regard to things like self-driving cars and smartphones, this apprehension has been replaced with comfort.

AGV manufacturers no longer have to assure their customers that the vehicles will improve safety and efficiency, as experience with similar technologies has removed previously ingrained concerns. Although this shift in temperament may seem unrelated to the rate at which organizations invest in AGVs, this fluctuation eliminated the final barrier preventing them from installing AGVs in their facilities.

The collapse of this wall has helped drive interest within the industry, translating into significant growth over the past few years. Industry 4.0 (also known as Industrie 4.0) has been another major propeller of increased interest in AGVs.

While the definition of this term can quickly become convoluted, it essentially boils down to implementing the Internet of Things (IoT) in a manufacturing environment.

In other words, those manufacturers who apply Industry 4.0 principles connect physical, automated equipment directly to software systems in order to enhance operations and to provide optimal efficiency. As a piece of automated equipment, AGVs logically fit into the Industry 4.0 puzzle. However, their importance becomes even more paramount when you consider the job function that they perform, namely moving materials. AGVs are the connective tissue that enables operations to work correctly. As a result, companies dedicated to pursuing Industry 4.0 initiatives have begun to consider AGVs a cornerstone of their strategy.

This has translated into real interest in the technology, how it can benefit organizations, and the best method of implementation. With all of this discussion regarding the increased interest in AGVs themselves, why do we anticipate that enterprise-wide deployments, in particular, will gain traction? The answer is simple — the numerous advantages that this type of plan presents versus traditional implementation methods.

By deploying AGVs across entire enterprises, companies can leverage repeatable processes and standardize the equipment and layout of every manufacturing, warehousing, and distribution facility in their control. These benefits also relate back to Industry 4.0, as this type of standardization creates connectivity across each of these plants and translates into more robust reporting and visibility into operations.

Aside from standardization and leveraging repeatability, companies can also optimize the labor savings generated from any type of automation. AGVs in an enterprise-wide program are built in bulk, lowering their overall cost. As such, the facilities in which these vehicles are introduced will achieve even more significant payback. This matters because labor rates continue to rise throughout the world.

Even in regions where these costs remain low, there is interest in implementing automation technologies in order to gain a competitive advantage over more developed countries. These benefits, plus many others that will go unmentioned, speak to the reasons why companies invest in enterprise-wide programs and are driving the interest in them. Based upon the number of partners that JBT already has and the number of enterprise program inquiries we are receiving, we anticipate that organizations will enter into these types of relationships even more quickly in the upcoming year. Interest in automated technology is higher than ever, and companies want to gain as much as possible from their investments.

This combination means that the growth of enterprise-wide AGV deployments will continue for the foreseeable future and will experience a significant upturn in 2018. ~

A Marketing Specialist at JBT, **Laura McConney** analyzes market behavior and tracks customers' buying journeys across the AGV industry.

Source: JBT

The Need to Overcome Interoperability and Labor Replacement Fears

Daniel Theobold

- Co-Founder of Vecna Robotics -

Until recently, large-scale use of mobile robots in industries was still aspirational for many reasons. Today, rising consumer expectations and intense global competition have created a sense of urgency to better understand how these solutions can create higher efficiency and solve many issues including sustainability.

Equally important, new market demand and cost-effectiveness of mobile robotic solutions are jumpstarting the mass adoption. While the true proliferation of robotic solutions is on the horizon, the issues of interoperability and market fears over job loss must be addressed and overcome.

Interoperability

One of the most significant barriers to the continued growth of robotic systems is a lack of interoperability within the industry. As robots become more prevalent, especially in manufacturing and warehousing, it is getting harder to manage the different systems — especially if they've been made by various companies.

Although collaborative environments that enable harmony between humans and robotic systems are important, it is much more critical to have robots that seamlessly integrate with other systems — hardware to hardware, hardware to software, and software to software. For the robotics industry to propel itself forward, it needs to be able to build on what others have already done — and to combine solutions to deliver more value.

This is a lesson learned from the computing industry; real adoption of computers in business did not occur until low-cost computers could run software written by a variety of companies. It was when they could easily connect to the rest of the world through the internet that the computer industry took off in a mass-consumer way. By building on each other's successes, we can rapidly advance the industry and therefore grow a bigger pie for the entire community. Computers have streamlined this process, as the feature that made them valuable is the ability to share information.

The inability for these systems to communicate with one another creates inefficient and potentially dangerous situations. For example, consider a robot's ability to use elevators. Right now, there is no standard for robot elevator use. Each company builds a custom solution, which could be problematic when multiple robots from different companies are operating in the same building. Given that there won't be a one-size-fits-all solution, interoperability is the only viable solution.

Because we are still in the early stages of market adoption, interoperability standards have not yet been set. This gives us an opportunity to be proactive as an industry and define standards that will help propel mass adoption of robots.

As an industry, we must open the lines of communication that allow us to address these types of issues. The need for a collaborative forum and defined standards is part of the mission of MassRobotics, an organization of which I'm a founder. Part of the reason we founded MassRobotics was to try and encourage this pre-competitive collaboration of robotics companies to build the industry together. We can share information and opportunities and work together to create reasonable standards of interoperability — a type of collaboration that is sorely lacking right now.

Collaboration within the community that adopts reasonable interoperability standards will accelerate the advancement of robotics.

Labor Replacement

Another issue that we as an industry must all address is the commonly surfaced debate of robotics replacing jobs. Complete lights-out automation will not be the norm for most industries. Instead, the industry should focus on robots that improve worker productivity and job satisfaction by automating the unwanted parts of jobs that humans don't find fulfilling.

With turnover rates as high as 300% for many manufacturing, shipping, and order-fulfillment jobs, Vecna Robotics is working with several employers that are struggling to find and keep the staff they need to meet customer expectations. By strategically deploying robots as helpers and promoting the human workers to more value-added tasks, everybody wins.

When people worry about losing their jobs, it cuts to the core of their confidence and positive outlook on life. That is part of what makes this such a sensitive and complex issue. If we aren't thoughtful about how we approach this issue, we risk doing more harm than good to the very people we are trying to help.

The short version of human history is simple: humans create technology, technology creates prosperity, prosperity is shared, human society thrives, and creates more technology. This is the "Prosperity Chain." Having more means I am likely to share more. Having more also allows me to spend a higher percentage of my time on things other than meeting my basic needs. Note that it is not necessary for me to care about other people for this to work, although I hope to show that if we do care, we will all do far better.

Will humanity need everyone "working" to meet our basic needs of food, clean water, clothing, and shelter? Not by a long shot. That was the humanity of days past. It takes only a small subset of society to create those basic needs. The rest of us are involved in what used to be generally considered "luxury" industries, such as education, science, medicine, travel, art, media, etc. Many don't consider these to be luxuries today, but for most of human history, these pursuits were only available to the elite. Prosperity has made them accessible to a broader circle of humanity, and that prosperity is attributed to technology.

My hope is that we can use our technology-created prosperity to evolve better socio-economic models over time that allow the debate to turn from "will everyone have a job?" to "does everyone have the opportunity to do something that they enjoy, and how can we align that with what we need to accomplish as a species?" Along the way, let's make sure to be aware of and take care of those that need a little extra help.

Robots and automation can create unprecedented levels of prosperity. The prosperity chain creates a virtuous feedback loop that allows for more prosperity and enables society to take care of more people than ever before. ~

Daniel Theobald co-founded Vecna in 1998 with the mission of empowering humanity through transformative technologies. Daniel is also Co-Founder and President of MassRobotics, an independent, nonprofit, industry consortium dedicated to fostering the global robotics community.

Source: Vecna Robotics

Redefining Growth

Fred van Beuningen

- Managing Partner of Chrysalix Venture Capital -

The last decades left us better housed, fed, and equipped through an economic growth miracle. That miracle came at a cost: environmental destruction puts our planet and companies at risk. The old economic model, with its resource overuse and product underuse, no longer works.

Cars and office buildings are underused, food is wasted, and non-renewable resources are depleted. At the same time, disruptive technologies that drive the intelligent systems revolution develop at breakneck speed and have large commercial benefits, but do they contribute to a new economic model where the economy prospers and nature thrives? Where renewable energy powers the economy, where everything that is used can be used again, and where less wasteful systems are designed for mobility, food, and housing? These will be important questions in 2018 and the years beyond.

The root cause of our wasteful economy is that we have developed efficient products but inefficient systems. The mobility system with extremely low car utilization and tank-to-wheel energy efficiency provides a good example. Land utilization and yearly deaths and injuries on roads further illustrate the structural waste in mobility. But waste is everywhere, not only in mobility, through low product utilization during product life, the short length of the product life, and the low reuse or recycle rates after the first-use cycle. The costs of our economic model have started to outpace the benefits so we live off the capital, not the dividends.

Whilst technological disruption provides part of the necessary system reset, it is important to define a set of principles and boundaries to make sure technology addresses relevant systems change. In their inspirational publication *A Good Disruption*, Stuchtey, Enkvist and Zumwinkel describe three building blocks for a positive system: abundant renewable energy, a cradle-to-cradle material bank, and a high-productivity industry built on circular and regenerative principles.

Intelligent Systems

Tipping-point technologies, like advanced sensors and IoT, combined with the availability of vast amount of data and computing power, have the potential to support system-level changes. At the same time, there is a significant spotlight on the ethical, legal, and safety aspects of future applications. In part, these will depend on how these technologies are deployed and what questions we use to inspire researchers and developers.

The mobility system illustrates the potential of positive disruption well. Ubiquitous connectivity and sensing enable us to measure things that were previously unmeasurable. Advances in data analytics enable us to build models to understand better how people move through cities. AI technology will help us move from descriptive models to predictive models and prescriptive decisions to optimize the flow of traffic in cities through new on-demand and multi-modal systems. AI technology can optimize traffic lights in real time, dispatch fleets of small electrical vehicles on demand, and address the "last mile" problem. Modal integration and sharing of transport will dramatically improve the utilization of vehicles and facilitate the inflow and outflow of cities, significantly reducing environmental impact and costs.

We need to become more intelligent about our resource use and redefine growth. Not as throughput, linear material flow and limited liability but as performance, circular material flow and internalized liabilities. This has important implications for business models, where customers become users and demand flexible access to the utility the product provides, and pay per use, facilitated by IoT, platforms, and devices.

This "performance economy" moves toward near-zero marginal cost, companies sell a service, retain ownership of the asset over the lifetime, and make a profit from the asset's use and end-of-life phase. Digital technology enables speed and scale of the most prominent circular business models such as resource recovery, sharing platforms, and product as a service.

These technologies provide the information and connections needed to maintain a relationship far beyond the point of sale. Such connections enhance remote visibility and control of assets, which are especially critical for the Product as a Service, Sharing Platforms, and Product Life Extension business models. By altering the way businesses and consumers interact with physical and digital assets and enabling dematerialization, digital technologies can transform value chains, so they are decoupled from the need for additional resources for growth. Core capabilities for companies will shift toward the ability to manage complex collaborative networks and innovation/product development. Customers will become more mobile and demanding, manufacturers will become retailers, and retailers will become digital communities.

Advanced AI capabilities can benefit public welfare, particularly in areas like safety, public health, education, and mobility. Public welfare, however, is a complex system, and improving, for example, a health or mobility system involves many interacting components. The ultimate grand challenge of AI for public welfare is to create tools that automatically and proactively identify problem causes, propose policy solutions, and predict consequences of those (potentially cross-issue) policies; to identify solutions that a person may not immediately come up with because of the ability of a systems model to look across domains and see linkages that no single individual could; to develop specific data sets to inform decisions based on system-level consequences.

Cities are the logical systems level at which to experiment with solutions that improve public welfare, and useful frameworks exist for how to build a great city. Best practices on how to combine the circular and smart vision as two reinforcing factors to create great cities become available. Circular thinking looks to extend the use cycle and utilization of assets, looping them into additional use cycles, and regenerate natural capital. The smart factor enables knowledge about the location, condition, and availability of the asset.

In the coming years, we will see more people translating the big smart and circular vision for cities into concrete programs and polices enabled by digital technology, distributed and clean energy, new business models, enlightened consumers, and governments acting as market makers. Parts of this transition will begin in 2018, but it may still be a number of years before it is commonplace. ~

Fred van Beuningen is an investor and industry executive in the clean tech sector. Fred is focused on helping to build intelligent and long-lasting technologies to push society forward in a clean and efficient manner. He is a Managing Partner of Crysalix Venture Capital.

THE FUTURIST INSTITUTE

 THE FUTURIST INSTITUTE

The Futurist Institute was founded in 2016 to help analysts become futurists by providing the content and context to take longer-term views on business opportunities, risk management, markets, and the economy. The Futurist Institute confers the Futurist and Long-Term Analyst™ (FLTA) designation and helps analysts become Certified Futurists™.

Current Courses

The Future of Work
The Future of Data
The Future of Finance
The Future of Transportation
Futurist Fundamentals

Future Courses

The Future of Energy
The Future of Leadership
The Future of Healthcare

Visit The Futurist Institute:

www.futuristinstitute.org

ABOUT THE EDITOR

Jason Schenker is the Chairman of The Futurist Institute and the President of Prestige Economics. He is also the world's top-ranked financial market futurist. Bloomberg News has ranked Mr. Schenker one of the most accurate forecasters in the world in 38 different categories since 2011, including #1 in the world in 23 categories for his forecasts of the Euro, the Pound, the Swiss Franc, crude oil prices, natural gas prices, gold prices, industrial metals prices, agricultural commodity prices, and U.S. non-farm payrolls.

Mr. Schenker has written four books that have been #1 Best Sellers on Amazon: *Commodity Prices 101*, *Recession-Proof*, *Electing Recess*ion, and *Jobs for Robots*. Mr. Schenker is also a columnist for *Bloomberg Prophets*. Mr. Schenker has appeared as a guest host on Bloomberg Television, as well as a guest on CNBC. He is frequently quoted in the press, including *The Wall Street Journal, The New York Times*, and *The Financial Times*.

Prior to founding Prestige Economics, Mr. Schenker worked for McKinsey & Company as a Risk Specialist, where he directed trading and risk initiatives on six continents. Before joining McKinsey, Mr. Schenker worked for Wachovia as an Economist.

Mr. Schenker holds a Master's in Applied Economics from UNC Greensboro, a Master's in Negotiation from CSU Dominguez Hills, a Master's in German from UNC Chapel Hill, and a Bachelor's in History and German from The University of Virginia. He also holds a certificate in FinTech from MIT, an executive certificate in Supply Chain Management from MIT, a graduate certificate in Professional Development from UNC, and an executive certificate in Negotiation from Harvard Law School. He is currently pursuing a certificate in Cybersecurity with NACD and Carnegie Mellon University. Mr. Schenker holds the professional designations CMT® (Chartered Market Technician), CVA® (Certified Valuation Analyst), ERP® (Energy Risk Professional), and CFP® (Certified Financial Planner).

Mr. Schenker is an instructor for LinkedIn Learning. His course on Financial Risk Management was released in October 2017. Additional courses will be forthcoming in 2018 and 2019.

Mr. Schenker is a member of the Texas Business Leadership Council, the only CEO-based public policy research organization in Texas, with a limited membership of 125 CEOs and Presidents. He is also a member of the 2018 Director class of the Texas Lyceum, a non-partisan, non-profit that fosters business and policy dialogue on important U.S. and Texas issues.

Mr. Schenker is an active executive in FinTech, as the founder of the foreign exchange FinTech startup Hedgefly, and as a Partner in the firm SpreadBot. He is also a member of the Central Texas Angel Network. Mr. Schenker is the Executive Director of the Texas Blockchain Association, and he is also a member of the National Association of Corporate Directors, as well as an NACD Board Governance Fellow.

In October 2016, Mr. Schenker founded The Futurist Institute to help analysts and economists become futurists through a training and certification program. He holds the Certified Futurist and Long-Term Analyst (FLTA) designation.

About Jason Schenker:

www.jasonschenker.com

About The Futurist Institute:

www.futuristinstitute.org

ABOUT THE PUBLISHER

Prestige Professional Publishing LLC was founded in 2011 to produce readable, insightful, and timely professional reference books. We are registered with the Library of Congress, and we are based in Austin, Texas.

Published Titles

Be The Shredder, Not The Shred

Commodity Prices 101

Electing Recession

Jobs for Robots

Robot-Proof Yourself

The Robot and Automation Almanac - 2018

Future Titles

Spikes: Growth Hacking Leadership

The Economics of Elections

The Brain Business

The Valuation Onion

DISCLAIMER

FROM THE PUBLISHER

The following disclaimer applies to any content in this book:

This book is commentary intended for general information use only and is not investment advice. Prestige Professional Publishing LLC does not make recommendations on any specific or general investments, investment types, asset classes, non-regulated markets (e.g., FX, commodities), specific equities, bonds, or other investment vehicles. Prestige Professional Publishing LLC does not guarantee the completeness or accuracy of analyses and statements in this book, nor does Prestige Professional Publishing LLC assume any liability for any losses that may result from the reliance by any person or entity on this information. Opinions, forecasts, and information are subject to change without notice. This book does not represent a solicitation or offer of financial or advisory services or products, and are market commentary intended and written for general information use only. This book does not constitute investment advice.

DISCLAIMER

FROM THE FUTURIST INSTITUTE

The following disclaimer applies to any content in this book:

This book is commentary intended for general information use only and is not investment advice. The Futurist Institute does not make recommendations on any specific or general investments, investment types, asset classes, non-regulated markets, specific equities, bonds, or other investment vehicles. The Futurist Institute does not guarantee the completeness or accuracy of analyses and statements in this book, nor does The Futurist Institute assume any liability for any losses that may result from the reliance by any person or entity on this information. Opinions, forecasts, and information are subject to change without notice. This book does not represent a solicitation or offer of financial or advisory services or products, and are market commentary intended and written for general information use only. This book does not constitute investment advice.

THE ROBOT AND AUTOMATION ALMANAC - 2018

COMPILED BY THE FUTURIST INSTITUTE

EDITED BY JASON SCHENKER

ISBN: 978-1-946197-03-0 *Paperback*
     978-1-946197-02-3 *Ebook*

Made in the USA
San Bernardino, CA
21 January 2018